CASE/ INTERNATIONAL

SHOP MANUAL

Information and Instructions

This shop manual contains several sections each covering a specific group of wheel type tractors. The Tab Index on the preceding page can be used to locate the section pertaining to each group of tractors. Each section contains the necessary specifications and the brief but terse procedural data needed by a mechanic when repairing a tractor on which he has had no previous actual experience.

Within each section, the material is arranged in a systematic order beginning with an index which is followed immediately by a Table of Condensed Service Specifications. These specifications include dimensions, fits, clearances and timing instructions. Next in order of arrangement is the procedures paragraphs.

In the procedures paragraphs, the order of presentation starts with the front axle system and steering and proceeding toward the rear axle. The last paragraphs are devoted to the power take-off and power lift systems. Interspersed where needed are additional tabular specifications pertaining to wear limits, torquing, etc.

HOW TO USE THE INDEX

Suppose you want to know the procedure for R&R (remove and reinstall) of the engine camshaft. Your first step is to look in the index under the main heading of ENGINE until you find the entry "Camshaft." Now read to the right where under the column covering the tractor you are repairing, you will find a number which indicates the beginning paragraph pertaining to the camshaft. To locate this wanted paragraph in the manual, turn the pages until the running index appearing on the top outside corner of each page contains the number you are seeking. In this paragraph you will find the information concerning the removal of the camshaft.

More information available at haynes.com
Phone: 805-498-6703

Haynes Group Limited
Haynes North America, Inc.

ISBN-10:0-87288-515-1
ISBN-13: 978-0-87288-515-8

Disclaimer

There are risks associated with automotive repairs. The ability to make repairs depends on the individual's skill, experience and proper tools. Individuals should act with due care and acknowledge and assume the risk of performing automotive repairs.

The purpose of this manual is to provide comprehensive, useful and accessible automotive repair information, to help you get the best value from your vehicle. However, this manual is not a substitute for a professional certified technician or mechanic.

This repair manual is produced by a third party and is not associated with an individual vehicle manufacturer. If there is any doubt or discrepancy between this manual and the owner's manual or the factory service manual, please refer to the factory service manual or seek assistance from a professional certified technician or mechanic.

Even though we have prepared this manual with extreme care and every attempt is made to ensure that the information in this manual is correct, neither the publisher nor the author can accept responsibility for loss, damage or injury caused by any errors in, or omissions from, the information given.

Common spark plug conditions

NORMAL
Symptoms: Brown to grayish-tan color and slight electrode wear. Correct heat range for engine and operating conditions.
Recommendation: When new spark plugs are installed, replace with plugs of the same heat range.

WORN
Symptoms: Rounded electrodes with a small amount of deposits on the firing end. Normal color. Causes hard starting in damp or cold weather and poor fuel economy.
Recommendation: Plugs have been left in the engine too long. Replace with new plugs of the same heat range. Follow the recommended maintenance schedule.

TOO HOT
Symptoms: Blistered, white insulator, eroded electrode and absence of deposits. Results in shortened plug life.
Recommendation: Check for the correct plug heat range, over-advanced ignition timing, lean fuel mixture, intake manifold vacuum leaks, sticking valves and insufficient engine cooling.

CARBON DEPOSITS
Symptoms: Dry sooty deposits indicate a rich mixture or weak ignition. Causes misfiring, hard starting and hesitation.
Recommendation: Make sure the plug has the correct heat range. Check for a clogged air filter or problem in the fuel system or engine management system. Also check for ignition system problems.

PREIGNITION
Symptoms: Melted electrodes. Insulators are white, but may be dirty due to misfiring or flying debris in the combustion chamber. Can lead to engine damage.
Recommendation: Check for the correct plug heat range, over-advanced ignition timing, lean fuel mixture, insufficient engine cooling and lack of lubrication.

ASH DEPOSITS
Symptoms: Light brown deposits encrusted on the side or center electrodes or both. Derived from oil and/or fuel additives. Excessive amounts may mask the spark, causing misfiring and hesitation during acceleration.
Recommendation: If excessive deposits accumulate over a short time or low mileage, install new valve guide seals to prevent seepage of oil into the combustion chambers. Also try changing gasoline brands.

HIGH SPEED GLAZING
Symptoms: Insulator has yellowish, glazed appearance. Indicates that combustion chamber temperatures have risen suddenly during hard acceleration. Normal deposits melt to form a conductive coating. Causes misfiring at high speeds.
Recommendation: Install new plugs. Consider using a colder plug if driving habits warrant.

OIL DEPOSITS
Symptoms: Oily coating caused by poor oil control. Oil is leaking past worn valve guides or piston rings into the combustion chamber. Causes hard starting, misfiring and hesitation.
Recommendation: Correct the mechanical condition with necessary repairs and install new plugs.

DETONATION
Symptoms: Insulators may be cracked or chipped. Improper gap setting techniques can also result in a fractured insulator tip. Can lead to piston damage.
Recommendation: Make sure the fuel anti-knock values meet engine requirements. Use care when setting the gaps on new plugs. Avoid lugging the engine.

GAP BRIDGING
Symptoms: Combustion deposits lodge between the electrodes. Heavy deposits accumulate and bridge the electrode gap. The plug ceases to fire, resulting in a dead cylinder.
Recommendation: Locate the faulty plug and remove the deposits from between the electrodes.

MECHANICAL DAMAGE
Symptoms: May be caused by a foreign object in the combustion chamber or the piston striking an incorrect reach (too long) plug. Causes a dead cylinder and could result in piston damage.
Recommendation: Repair the mechanical damage. Remove the foreign object from the engine and/or install the correct reach plug.

SHOP MANUAL
CASE/INTERNATIONAL

MODELS 5120, 5130 & 5140

IDENTIFICATION

Tractor model number and identification serial number are located on a plate on lower left side of instrument panel cover on tractors equipped with a cab or lower left front side of operator's platform on tractors equipped with a ROPS frame. Engine serial number is located on a plate on left side of timing gear housing. Transmission serial number is located on a plate on right side of transmission housing. On models equipped with mechanical front-wheel-drive axle, the axle serial number is located on a plate on right rear side of front axle housing.

INDEX (By Starting Paragraph)

INDEX (CONT.)

INDEX (CONT.)

INDEX (CONT.)

CONDENSED SERVICE DATA

	Models 5120	5130	5140
GENERAL			
Engine Make		Own	
Number of Cylinders	4	6	6
Bore	102 mm (4.02 in.)	102 mm (4.02 in.)	102 mm (4.02 in.)
Stroke	120 mm (4.72 in.)	120 mm (4.72 in.)	120 mm (4.72 in.)
Displacement	3.92 L (239 cu. in.)	5.88 L (359 cu. in.)	5.88 L (359 cu. in.)
Compression Ratio	16.5:1	17:1	17.5:1
Electrical System		12 volts, negative ground	
Speed Selection		16 forward, 12 reverse	
TUNE-UP			
Firing Order	1-3-4-2	1-5-3-6-2-4	1-5-3-6-2-4
Rocker Arm-to-Valve Clearance (Cold)—			
Intake		0.25 mm (0.010 in.)	
Exhaust		0.50 mm (0.020 in.)	
Injection Timing	Paragraph 61	Paragraph 60	Paragraph 61
Injector Opening Pressure		24,500-25,305 kPa (3553-3670 psi)	
Engine Low Idle		900 rpm	
Engine High Idle (No-Load)		2400 rpm	

CONDENSED SERVICE DATA (CONT.)

	Models		
	5120	**5130**	**5140**

TUNE-UP (Cont.)
Engine Rated Speed (Loaded)................ _____2200 rpm_____

SIZES AND CLEARANCES
Crankshaft Main Journal Diameter........... _____ 82.987-83.013 mm _____
    (3.2672-3.2682 in.)
Main Bearing Diametral Clearance........... _____ 0.041-0.119 mm _____
    (0.0016-0.0046 in.)
Crankshaft End Play....................... _____ 0.137-0.264 mm _____
    (0.005-0.010 in.)
Crankshaft Crankpin Diameter.............. _____68.987-69.013 mm _____
    (2.706-2.717 in.)
Rod Bearing Diametral Clearance............ _____ 0.038-0.116 mm _____
    (0.0015-0.0045 in.)
Connecting Rod Side Clearance.............. _____ 0.1-0.3 mm_____
    (0.004-0.012 in.)
Camshaft Journal Diameter................. _____53.987-54.013 mm _____
    (2.125-2.126 in.)
Camshaft Diametral Clearance.............. _____ 0.076-0.15 mm_____
    (0.003-0.006 in.)
Camshaft End Play........................ _____ 0.13-0.47 mm _____
    (0.005-0.018 in.)
Cylinder Bore............................. _____ 102.00-102.04 mm _____
    (4.016-4.017 in.)
Piston Skirt Diameter...................... _____ 101.873-101.887 mm _____
    (4.010-4.011 in.)
Piston Pin Diameter .... _____ 39.997-40.003 mm _____
    (1.5747-1.5749 in.)
Piston Pin Clearance in Piston.............. _____0.003-0.015 mm_____
    (0.0001-0.0006 in.)
Piston Pin Clearance in Rod................. _____ 0.051-076 mm _____
    (0.002-0.003 in.)
Valve Stem Diameter—
  Intake and Exhaust...................... _____ 7.960-7.980 mm _____
    (0.313-0.314 in.)
Valve Face Angle—
  Intake................................ _____29°_____
  Exhaust.............................. _____44°_____
Valve Seat Angle—
  Intake................................ _____ 30°_____
  Exhaust.............................. _____ 45°_____

CAPACITIES
Cooling System—

With Cab................................	17.6 L	20.8 L	20.8 L
	(18.6 U.S. Qt.)	(22 U.S. Qt.)	(22 U.S. Qt.)
Without Cab............................	15.7 L	18.9 L	18.9 L
	(16.6 U.S. Qt.)	(20 U.S. Qt.)	(20 U.S. Qt.)

Crankcase—

With Filter	10.4 L	15 L	15 L
	(10.9 U.S. Qt.)	(15.8 U.S. Qt.)	(15.8 U.S. Qt.)

CONDENSED SERVICE DATA (CONT.)

	Models		
	5120	5130	5140
CAPACITIES (Cont.)			
Without Filter........................	9.5 L	14.3 L	14.3 L
	(10 U.S. Qt.)	(15.1 U.S. Qt.)	(15.1 U.S. Qt.)
Transmission/Hydraulic System..............		76 L	
		(20 U.S. Gal.)	
Front Drive Axle—			
Differential		6.5 L	
		(6.8 U.S. Qt.)	
Planetaries.........................		0.9 L	
		(0.95 U.S. Qt.)	
Fuel Tank (Std.)........................	135 L	130 L	130 L
	(35.6 U.S. Gal.)	(34.3 U.S. Gal.)	(34.3 U.S. Gal.)
Fuel Tank (Opt.)........................		170 L	
		(45 U.S. Gal.)	

DUAL DIMENSIONS

This service manual provides specifications in both the Metric (SI) and U.S. Customary systems of measurement. The first specification is given in the measuring system perceived by us to be the preferred system when servicing a particular component; the second specification (given in parentheses) is the converted measurement. For instance, a specification of "0.28 mm (0.011 inch)" would indicate that we feel the preferred measurement, in this instance, is the metric system of measurement and the U.S. system equivalent of 0.28 mm is 0.011 inch.

FRONT AXLE (TWO-WHEEL DRIVE)

FRONT WHEEL BEARINGS

All Models

1. Refer to Fig. 1 for typical wheel hub and bearing assembly.

The tapered inner and outer roller bearings (4 and 10) are not interchangeable. Clean and inspect bearing cups and cones and renew as necessary.

Install inner bearing cone (4) on steering knuckle spindle. Install bearing cups (5 and 9) in hub. Install oil seal (3) and clamping ring (2) in inner bore of hub. Pack bearing cones and fill hub with No. 2 lithium base grease. Install hub on steering knuckle, then install outer bearing cone (10), washer (11) and nut (12).

To adjust the bearings, tighten nut to a torque of 100 N.m (74 ft.-lbs.) while rotating the hub. Then, back nut off until next cotter pin slot is aligned. Install cotter pin. Install and tighten dust cover (13).

Install wheel and tire, and tighten lug nuts (8) to a torque of 300 N.m (220 ft. lbs.).

STEERING KNUCKLES

All models

2. REMOVE AND REINSTALL. To remove either steering knuckle, raise front of tractor and place a stand under axle extension. Remove wheel and tire assembly. Disconnect tie rod ball joint from steering arm (8—Fig. 2) and steering cylinder ball joint (13) from right steering arm. Support hub and steering knuckle and remove steering arm clamping bolt. Lift off steering arm and shims (9). Lower steering knuckle from axle extension (12).

Remove felt seal (10) and upper bushing (11) from axle extension. Remove lower bushing (17) and thrust bearing (18) from steering knuckle. Split bushings (11 and 17) are not a tight fit in axle extensions. If necessary, remove hub and wheel bearings as outlined in paragraph 1.

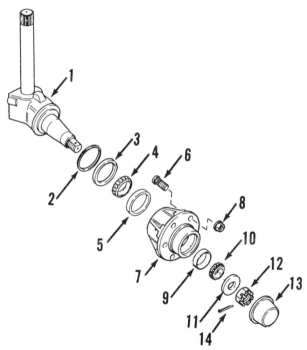

Fig. 1—Exploded view of steering knuckle, front wheel hub and bearings used on all two-wheel-drive tractors.

1. Steering knuckle	8. Lug nut
2. Clamping ring	9. Bearing cup
3. Oil seal	10. Bearing cone
4. Bearing cone	11. Washer
5. Bearing cup	12. Nut
6. Lug bolt	13. Dust cap
7. Hub	14. Cotter pin

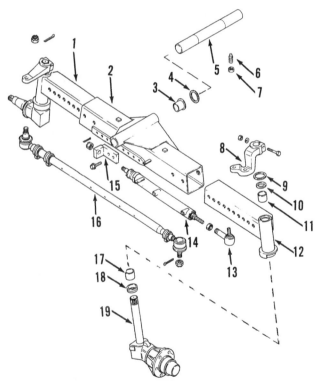

Fig. 2—Partially exploded view of front axle assembly used on all two-wheel-drive tractors.

1. Axle extension assy. (L.H.)	
2. Axle main mamber	11. Bushing
3. Pivot bushing (2)	12. Axle extension (R.H.)
4. Shim	13. Ball joint
5. Pivot pin	14. Steering cylinder
6. Allen set screw	15. Bracket
7. Locknut	16. Tie rod assy.
8. Steering arm (R.H.)	17. Bushing
9. Shim	18. Thrust bearing
10. Felt seal	19. Steering knuckle (R.H.)

Clean and inspect all parts and renew as required. Reinstall by reversing the removal procedure. Add or remove shims (9) to obtain 0.1 mm (0.004 in.) end play for the steering knuckle. Tighten steering arm clamping bolt to a torque of 280 N·m (206 ft.-lbs.).

TIE ROD AND TOE-IN

All Models

3. Disassembly of the tie rod is obvious after an examination of the unit and reference to Fig. 3.

Front wheel toe-in should be 2-4 mm (0.080-0.160 in.). To check toe-in, measure the distance between the front wheel rims at the front and at the rear of the wheel rims at hub height. The measurement at

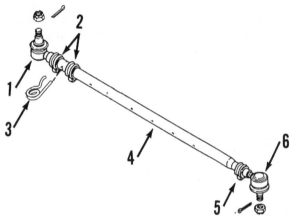

Fig. 3—Tie rod used on all two-wheel-drive tractors.

1. Left tie rod end & inner rod
2. Clamps
3. Spring pin
4. Outer tube
5. Clamp
6. Right tie rod end

the front must be 2-4 mm (0.080-0.160 in.) less than measurement at the rear.

To adjust toe-in, loosen clamps (2—Fig. 3) and remove spring pin (3). Loosen clamp (5) and rotate outer tube (4) as required. Reinstall spring pin (3) and tighten clamps.

AXLE MAIN MEMBER AND PIVOT PIN

All Models

4. REMOVE AND REINSTALL. To remove the axle main member (2—Fig. 2), engage park brake and securely block rear wheels. Remove front weights and bracket, if so equipped. Disconnect steering tube clamp, then disconnect steering tubes from steering cylinder (14). Raise and support front of tractor and place a floor jack under axle main member. Remove steering cylinder and tie rod. Unbolt and remove axle extensions (1 and 12) with steering knuckles, wheel and hub assemblies. Remove protective plug in front of pivot pin (5). Loosen locknut (7) and Allen set screw (6). Using a slide hammer and adapter (CAS-2014), pull pivot pin forward from axle main member and front support. Remove shims (4) from in front and rear of axle main member as pivot pin is removed. Lower axle main member from tractor.

Inspect front and rear pivot bushings (3) and pivot pin (5) for wear and renew as necessary. When reinstalling, reverse the removal procedures, keeping the following points in mind. Place one 0.5 mm (0.020 in.) thick shim at the rear of the axle and install enough shims at front of axle to obtain the required 0.1-0.3 mm (0.004-0.012 in.) end play. Tighten the Allen set screw and locknut (6 and 7) to a torque of 150 N·m (110 ft.-lbs.).

FRONT DRIVE AXLE

All models are available with a front-wheel drive axle. Front drive axle can be engaged or disengaged with tractor on the go. Front drive axle is spring-clutch engaged and hydraulically disengaged. Tractor must be split between speed transmission and range transmission for access to the front-wheel drive clutch. Front drive axle differential unit is equipped with a limited slip differential.

DRIVE AXLE ASSEMBLY

All Models So Equipped

5. REMOVE AND REINSTALL. To remove the front drive axle assembly (7—Fig. 4), apply park brake and block rear wheels securely. If so equipped, remove front weights and weight bracket. Loosen front wheel lug nuts. Unbolt and remove front drive shaft and shield as outlined in paragraph 9. Disconnect steering hoses at cylinder. Cap or plug all openings immediately to prevent entrance of dirt into system. Place a floor jack under front axle and raise front of tractor. Support front of tractor with jack stands under side rails. Remove front wheel and tire assemblies. Remove locknut (4) and Allen set screw (3). Remove protective plug (1), then using a slide hammer and adapter (CAS-2014), pull pivot pin (2) from axle and front support. Remove shims (5) as pin is removed. Lower axle assembly and remove from under tractor.

Bushings (6) can now be removed using a suitable drift punch. Drive new bushings into bore of axle main member until flush to 1.6 mm (¹⁄₁₆ in.) below outer surface.

When reassembling, install axle assembly without shims (5). With axle main member pushed rearward against front support, measure distance between front of axle and front support. Subtract 0.1-0.3 mm (0.004-0.012 in.) from this measurement for total shim pack to be installed. Divide this shim pack and install half in front and half at rear of axle main member.

The balance of installation is the reverse order of removal, keeping the following points in mind. Tighten the Allen set screw and locknut (3 and 4) to a torque of 150 N·m (110 ft.-lbs.). Tighten wheel lug nuts to a torque of 185-220 N·m (137-162 ft.-lbs.). Tighten front weight bracket bolts to a torque of 610-730 N·m (450-540 ft.-lbs.).

Refer to paragraph 9 for installation of front drive shaft and shield.

WHEEL HUB AND PLANETARY

All Models So Equipped

6. R&R AND OVERHAUL. To remove either front wheel hub and planetary, loosen wheel lug nuts, then raise and support axle housing. Remove front wheel assembly. Rotate hub until oil plug (25—Fig. 5) is at bottom position. Remove plug and drain oil. Remove two screws (26) and locate the two special wheel studs (with collar). Install two nuts on each stud and lock them together. Unscrew and remove the special studs. Using large screwdrivers or pry bars, separate planetary carrier (20) from wheel hub (3). Remove three screws (29) and lift off cover (28). Number the carrier, planetary gears (16) and shafts (22) for aid in correct reassembly. Drive out roll pins (21) as shown in Fig. 6. Place outer edge of planetary carrier on blocks, then drive out shafts (22—Fig. 5). Remove and discard "O" rings (24). Remove planetary gears (16), washers (15 and 18) and bearing rollers (17).

Remove snap ring (14), sun gear (13), spacer (12) and thrust washer (11). Remove the eight cap screws (8) from ring gear carrier (6). Use four of the cap screws as jack screws in threaded holes in ring gear carrier and remove ring gear and carrier assembly. Remove retaining rings (5 and 9), then separate ring gear (10) from carrier (6). Remove outer bearing cone (4), wheel hub (3) and inner bearing cone (2). Remove oil seal (1) from hub, then drive bearing cups (2 and 4) from hub if necessary.

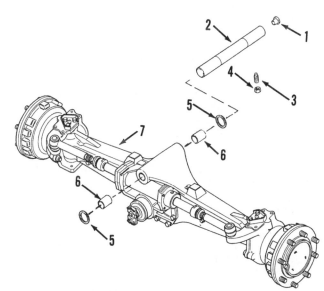

Fig. 4—View of front drive axle assembly removed from tractor.

1. Protective plug
2. Pivot pin
3. Allen set screw
4. Locknut
5. Shims
6. Pivot bushings
7. Axle assy.

Clean and inspect all parts and renew any showing excessive wear or other damage. Reinstall by reversing removal procedures keeping the following points in mind: Use new oil seal (1) and "O" rings (24). Tighten ring gear carrier cap screws (8) to a torque of 200 N·m (162 ft.-lbs.). When installing planetary carrier (20) on wheel hub (3), apply Activator (CAS-345153) to face of planetary carrier and Sealant (CAS-345152) to face of hub. Within five minutes after assembly, tighten the two special wheel studs to a torque of 70 N·m (51 ft.-lbs.) and the two countersunk screws (26) to a torque of 80 N·m (59 ft.-lbs.). Install cover (28) and tighten the three screws (29) to a torque of 34 N·m (25 ft.-lbs.).

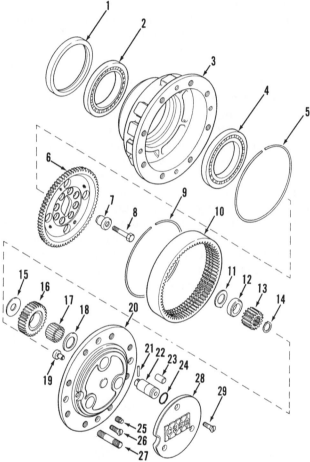

Fig. 5—Exploded view of front drive axle wheel hub and planetary assembly.

1. Oil seal
2. Bearing cup & cone
3. Wheel hub
4. Bearing cup & cone
5. Retaining ring
6. Ring gear carrier
7. Bushing
8. Cap screw
9. Retaining ring
10. Ring gear
11. Thrust washer
12. Spacer
13. Sun gear
14. Snap ring
15. Washer
16. Planetary gear (3)
17. Bearing rollers
18. Washer
19. Thrust plug
20. Planet gear carrier
21. Roll pin
22. Shaft (3)
23. Dowel pin
24. "O" ring
25. Oil plug
26. Screw
27. Wheel stud
28. Cover
29. Screw

Rotate hub until the word CASE is horizontal, then fill the unit to level plug opening with SAE 85W-140 gear lubricant. Capacity is approximately 1.0 liter (1.06 U.S. qt.). Install plug (25). Install wheel and tighten lug nuts to 300 N·m (220 ft.-lbs.) torque.

Repeat operation on opposite wheel hub and planetary if necessary.

PIVOT HOUSING AND AXLE SHAFT

All Models So Equipped

7. REMOVE AND REINSTALL. To remove the pivot housing (6—Fig. 7), first remove the wheel hub and planetary as outlined in paragraph 6. Disconnect tie rod end from steering arm. Remove cap screws (1), then remove pivot pins (3) with shims (4) and bearing cones (5). Lift off pivot housing (6). Remove oil seal (11) and, if necessary, remove bushing (12) from pivot housing. Remove oil seals (8) and bearing cups (9). Use a suitable puller to remove bearing cones (5) from pivot pins.

Withdraw axle shaft assembly (7). Remove oil seal (5—Fig. 8) and, if necessary, remove bushing (6).

Clean and inspect all parts and renew any showing excessive wear or other damage. "U" joint cross and bearings (7) are available only as a unit. Renewal of "U" joints is obvious after examination of the assembly and reference to Fig. 8.

When reassembling, use all new oil seals. Install seal (5—Fig. 8) in axle housing (2), then carefully install axle shaft assembly (7—Fig. 7). Install oil seal (11) in pivot housing (6). Install bearing cups (9) and oil seals (8) in axle housing (10). Carefully install pivot housing over outer axle shaft and axle housing. Install lower pivot pin (3) with bearing cone (5), but without shims (4). Tighten lower cap screws (1) to a torque of 83 N·m (60 ft.-lbs.). Install upper pivot pin (3) with bearing cone (5), but without shims (4). Use a jack under lower pivot pin to hold pivot housing upward. While holding down on pivot pin, use a feeler

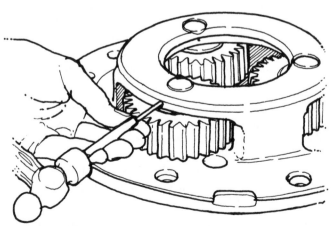

Fig. 6—When disassembling planetary, drive out shaft roll pins as shown.

gauge to measure distance between pivot housing and flange of pivot pin. Select a shim pack 0.1-0.2 mm (0.004-0.008 in.) less than the measurement. Divide the selected shim pack equally and install half on top pivot pin and half on bottom pivot pin. Shims are available in thicknesses of 0.10, 0.19 and 0.35 mm (0.004, 0.007 and 0.014 in.). Tighten top and bottom cap screws (1) to a torque of 137 N·m (100 ft.-lbs.).

The balance of reassembly is the reverse order of disassembly. Lubricate pivot pin bearings with CASE 251 H EP or multipurpose lithium grease. Connect tie rod end to steering arm and tighten nut to a torque of 230 N·m (170 ft.-lbs.).

DIFFERENTIAL AND BEVEL GEARS

All Models So Equipped

8. R&R AND OVERHAUL. To remove the differential assembly, first remove drain plug and drain oil from axle housing. Remove front drive axle assembly as outlined in paragraph 5 and place assembly on axle stands. Disconnect tie rods from steering arms. Remove pivot pins (2—Fig. 9), then lift off wheel hub,

planetary, pivot housing and axle shaft assembly (3). Repeat operation on opposite side. Remove tie rods (4 and 9) from steering cylinder (7). Unbolt bracket (8), then unbolt and remove steering cylinder. Remove retaining bolts and lift out pinion housing and differential assembly (5).

Refer to Fig. 10 and unbolt and remove locks (2). Loosen cap screws (1) and back off adjusting rings (4 and 6). Remove cap screws (1), bearing caps (3) and adjusting rings. Lift out ring gear and differential assembly (5) and lay aside for later disassembly.

Remove snap ring (21), front yoke (20), "O" ring (19), washer (18) and oil seal (17). Using special socket wrench (CAS-1885), remove nut (16) and tab washer (15). Use a soft hammer and drive pinion shaft (7) from housing (11). Remove bearing cone (14) and spacer (12), then press bearing cone (9) from

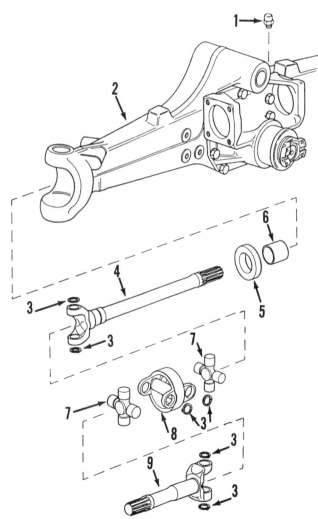

Fig. 8—Exploded view of axle shaft assembly.

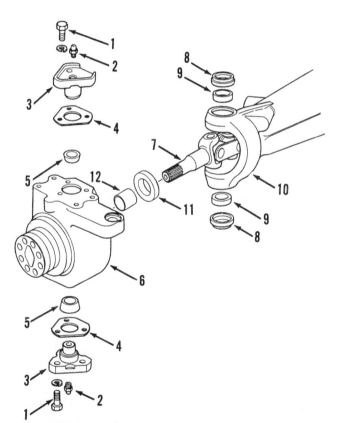

Fig. 7—Exploded view of pivot housing and components used on front drive axle.

1. Cap screws	7. Axle shaft assy.
2. Lube fittings	8. Oil seals
3. Pivot pins	9. Bearing cups
4. Shims	10. Axle housing
5. Bearing cones	11. Oil seal
6. Pivot housing	12. Bushing

1. Breather	6. Bushing
2. Axle housing	7. "U" joint cross
3. Snap rings	& bearings
4. Axle shaft (inner)	8. Center yoke
5. Oil seal	9. Axle shaft (outer)

pinion shaft (7) and remove shim (8). Remove bearing cups (10 and 13) from housing.

To disassemble differential assembly (5), refer to Fig. 11 and remove bearing cups (1) and bearing cones (2). Place match marks across differential housing halves and ring gear. Remove cap screws (14), and remove ring gear (13) and right housing half (12). Remove separator plates (4), friction plates (5), thrust plate (6) and right side gear (7), then remove cross shafts (10), differential pinions (9) and thrust washers (8). Remove left side gear (7) and thrust plate (6), separator plates (4) and friction plates (5) from left housing half (3).

Clean and inspect all parts and renew any showing excessive wear or other damage. If ring gear (13) or drive pinion (7—Fig. 10) are faulty, both must be renewed as they are only available as a matched set. Differential plates should be within the following thickness ranges:

Separator plates . . . . 1.47-1.55 mm
                                              (0.058-0.061 in.)
Friction plates . . . . 1.57-1.63 mm
                                              (0.062-0.064 in.)
Thrust plates . . . . 2.77-2.83 mm
                                              (0.109-0.111 in.)

To install and adjust bevel drive pinion position, proceed as follows: Install bearing cups (10 and 13—Fig. 10) in housing (11). Install bearing cone (4—Fig. 12) over rod end (3) of special tool (CAS-1839) and install in front of housing. Install bearing cone (8) and handle (9) in rear of housing. Tighten special tool by hand until bearing cones are seated in bearing cups (all end play removed). Install tube (2) of special tool and clamp in place with bearing caps (1). Measure and record distance between tube (2) and gauge block (3). Add this measurement to the pinion height con-

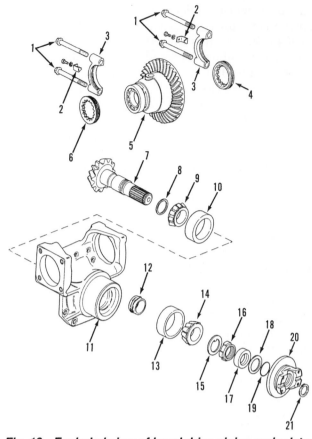

Fig. 10—Exploded view of bevel drive pinion and related components. Ring gear and bevel drive pinion are available only as a matched set.

1. Cap screws
2. Locks
3. Bearing caps
4. Adjusting ring (R.H.)
5. Ring gear & differential assy.
6. Adjusting ring (L.H.)
7. Bevel pinion shaft
8. Shim
9. Bearing cone
10. Bearing cup
11. Pinion housing
12. Collapsible spacer
13. Bearing cup
14. Bearing cone
15. Washer
16. Nut
17. Oil seal
18. Washer
19. "O" ring
20. Front yoke
21. Snap ring

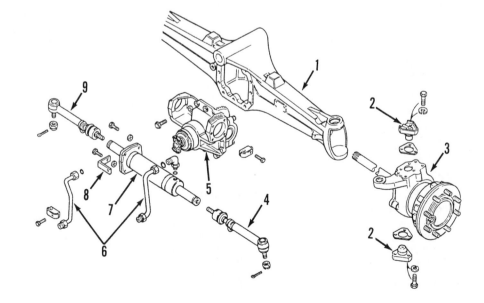

Fig. 9—Differential assembly removed from front drive axle housing.

1. Axle housing
2. Pivot pins
3. Wheel hub, planetary, pivot housing & axle shaft assy.
4. Right tie rod assy.
5. Differential assy.
6. Steering lines
7. Steering cylinder
8. Bracket
9. Left tie rod assy.

stant, 109.75 mm (4.32 in.). Then, subtract the measurement etched on pinion shaft in millimeters from the previous sum to obtain the correct thickness of shims (8—Fig. 10) to be installed. Remove the special tool (CAS-1839) and bearing cones. Shims are available in ten thicknesses from 2.5-3.4 mm (0.098 to 0.134 in.).

Install the previously determined shim pack (8) on pinion shaft (7). Heat bearing (9) to a temperature of 80-90° C (175-195° F) and install on pinion shaft. Install collapsible spacer (12) on shaft, then install pinion shaft assembly in housing (11). Heat and install bearing cone (14) and install tab washer (15) and a new nut (16). Tighten nut only until shaft end play is removed. Use a cord wrapped around splined end of pinion shaft and a spring scale to measure the force required to rotate the pinion shaft. Bearing preload is correct when a spring scale reading of 92-137 N (20.7-30.8 lbs.) is obtained. Tighten pinion shaft nut until desired spring scale reading (bearing preload) is obtained. When preload is correct, use a hammer

and small punch to stake edge of nut into slot on pinion shaft.

> **NOTE: If the pinion shaft nut is overtightened and the bearing preload exceeds 137 N (30.8 lbs.), disassemble the pinion shaft and install a new collapsible spacer (12). Then, repeat the bearing preload adjustment procedure.**

Using Fig. 11 as a guide, reassemble ring gear and differential assembly. Soak all plates (4, 5 and 6) in clean SAE 85W-140 gear oil for 30 minutes prior to installing. Friction side of thrust plate (6) must face separator plate. Align match marks (applied during disassembly) on differential halves (3 and 12) and ring gear (13). Apply Loctite 270 to the cap screws (14), install and tighten cap screws to a torque of 78 N·m (57 ft.-lbs.). Heat bearing cones (2) to a temperature of 80-90° C (175-195° F), then install them on differential housing. Place bearing cups (1) on bearing cones (2) and install ring gear and differential assembly (5—Fig. 10) on pinion housing (11). Make sure bearing cups are seated and that ring gear is on drain plug side of housing. Install adjusting rings (4 and 6), bearing caps (3) and cap screws (1). Turn adjusting ring (4) until ring gear just touches pinion gear, then tighten adjusting ring (6) until ring becomes difficult to turn with special tool (CAS-1840B). Use a dial indicator against a ring gear tooth and check backlash between ring gear and pinion. Turn adjusting rings (4 and 6) equal amounts until 0.17-0.24 mm (0.007-0.009 in.) backlash is obtained. Check backlash at four equal positions around ring gear.

With backlash correctly set, wrap a cord around pinion shaft splines and attach a spring scale. Measure the force required to rotate the pinion shaft.

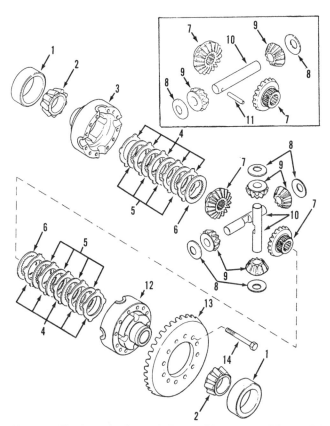

Fig. 11—Exploded view of front drive axle differential assembly. Two pinion differential (inset) is used on Model 5120.

1. Bearing cups	8. Thrust washers
2. Bearing cones	9. Differential pinions
3. Differential housing	10. Cross-pins
(left half)	11. Pin
4. Separator plates	12. Differential housing
5. Friction plates	(right half)
6. Thrust plates	13. Ring gear
7. Side gears	14. Cap screw

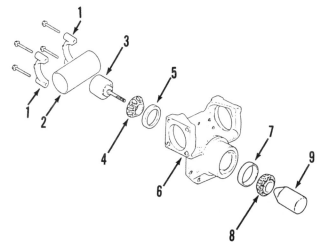

Fig. 12—View showing special tool (CAS-1839) used to determine drive pinion shaft shim thickness. Refer to text.

1. Bearing caps	
2. Gauge tube	
3. Gauge block	6. Pinion housing
4. Bearing cone	7. Bearing cup
5. Bearing cup	8. Bearing cone
	9. Tool handle

Spring scale reading (bearing preload) should be 31-46 N (7-10 lbs.) above the pinion bearing preload of 92-137 N (20.7-30.8 lbs.) previously set. Tighten both adjusting rings an equal amount to get the recommended preload. Recheck backlash and readjust if necessary. Tighten cap screws (1) to 266 N·m (196 ft.-lbs) torque. Install locks (2) and secure with cap screws tightened to 10 N·m (7 ft.-lbs.) torque.

Install thrust washer (18) and new "O" ring (19) over end of pinion shaft. Apply Loctite 515 to metal cover of new oil seal (17) and install seal until seated in pinion housing. Install front yoke (20) and snap ring (21).

When reinstalling the bevel gears and differential assembly, apply Loctite 515 to pinion housing mounting face and tighten retaining bolts to a torque of 120 N·m (88 ft.-lbs.) on axles equipped with bolts and washers or 169 N·m (125 ft.-lbs.) on axles equipped with self-locking bolts.

Install steering cylinder (7—Fig. 9) and bracket (8). Tighten retaining cap screws to a torque of 90 N·m

(66 ft.-lbs.). Install tie rod assemblies (4 and 9) and tighten into cylinder rod to a torque of 300 N·m (221 ft.-lbs.). Install the wheel hub, planetary, pivot housing and axle shaft assemblies (3) by reversing the removal procedures. Tighten pivot pin cap screws to a torque of 137 N·m (101 ft.-lbs.). Connect tie rod ends to steering arms and tighten nuts to 230 N·m (170 ft.-lbs.) torque. Reinstall drive axle assembly as outlined in paragraph 5. Fill axle housing with SAE 85W-140 gear oil to level plug opening. Capacity is approximately 6.0 liters (6.3 U.S. qt.). Use new "O" rings and reconnect steering lines. The balance of reassembly is the reverse of disassembly.

FRONT DRIVE SHAFT

All Models So Equipped

9. R&R AND OVERHAUL. To remove front drive shaft, block rear wheels securely. Refer to Fig. 13 and unbolt and remove shield (5) with engine oil drain plug rubber protector (6). Unbolt "U" joints from front and rear yokes and remove drive shaft assembly. Unbolt "U" joints from drive shaft. Place mark on rear yoke (2) and transmission housing for aid in reassembly. Pull rear yoke from transmission. Remove oil seal (1) from rear yoke. If necessary, remove snap ring (21—Fig. 10) and remove front yoke (20).

Clean and inspect all parts for excessive wear or other damage and renew as necessary. Install new oil seal and reassemble drive shaft assembly. Reinstall by reversing removal procedure. Tighten bolts in "U" joints to a torque of 39-44 N·m (29-32 ft.-lbs.). Install shield with engine oil drain plug rubber protector.

TIE RODS AND TOE-IN

All Models So Equipped

10. ADJUSTMENT. To check and adjust the toe-in, refer to Fig. 14 and measure distance "A" and

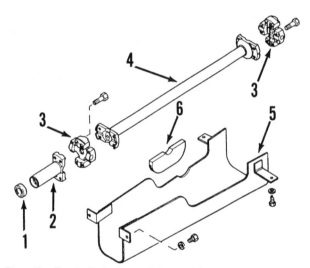

Fig. 13—Exploded view of front drive shaft and shield. Refer to Fig. 10 for view of front yoke.

1. Oil seal	4. Drive shaft
2. Rear yoke	5. Shield
3. "U" joints	6. Oil drain rubber protector

Fig. 14—Top view of front drive axle showing tie rod adjustment clamps. Refer to text.

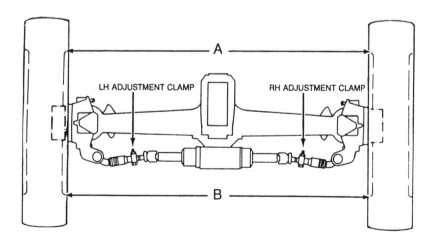

distance "B" from rim to rim at hub height. Distance "A" at front of wheel rim should be equal to or up to 5.0 mm (0.197 in.) less than distance "B" at rear of wheel rim.

If toe-in is incorrect, loosen adjustment clamps, then rotate steering cylinder piston rod as required to increase or decrease toe-in. Retighten tie rod clamps.

FRONT DRIVE AXLE CLUTCH

All Models So Equipped

11. CLUTCH TORQUE TEST. To check the front drive clutch slip torque, remove front drive shaft and shield as outlined in paragraph 9. Install special adapter (CAS-2131) to rear yoke (2—Fig. 13) using "U" joint bolts. Tighten bolts to 29-32 ft.-lbs. (39-44 N·m). Have an assistant apply the service brakes and, using a torque multiplier (OEM-6107) and torque meter (OEM-6481), check torque required to turn the output shaft. Torque should be as follows:

New clutch . . . . 2100 N·m
(1550 ft.-lbs.)
Used clutch . . . . 1260-1540 N·m
(930-1135 ft.-lbs.)
Clutch with worn plates . . . . 1000 N·m or less
(738 ft.-lbs. or less)

If clutch slips at a torque of 1000 N·m (738 ft.-lbs.) or less, remove and overhaul clutch as outlined in paragraph 12.

12. R&R AND OVERHAUL. To remove the clutch, tractor must be split between the speed and range transmissions as follows: Block front wheels securely. Install wood wedges between front axle housing and front support. Disconnect battery cables and drain transmission oil. Unbolt and remove seat assembly. Identify electrical connections and disconnect shuttle valve solenoids, pressure switches and ground leads. Disconnect and cap hydraulic lines as required. Disconnect the inching clutch cable and the park brake cable. Unbolt and remove left step assembly. Drain fuel tank and disconnect breather hose, return hose and fuel supply hose. Unbolt and remove fuel tank. Remove the bottom three bolts from transmission. Install transmission splitting rail (CAS-10853) under range transmission and install a suitable safety stand under speed transmission. Remove the right rear wheel. Install cab stands (CAS-3389) under cab frame. Disconnect pto cable, then remove cable and switch from the bracket. Unbolt and remove trailer electrical socket. Disconnect ground speed sensor connector and disconnect remote valve control rods. Disconnect and cap brake hoses. Unbolt and remove the shield, then disconnect the

steering pressure switch connector, draft control solenoid connector and range switch connectors. Disconnect range selector rods and disconnect front-wheel drive solenoid connector. If so equipped, disconnect creeper transmission cable. Disconnect and cap hydraulic tubes from front-wheel drive valve and compensator valve. Raise the hood, remove exhaust extension and rear hood panel. Disconnect ground cables on both sides of cab. Remove rear cab mounting nuts and loosen front cab mounting bolts. Using the three cab support stands, raise cab. Remove speed to range transmission mounting bolts and carefully separate range transmission from speed transmission.

Remove clutch assembly from front of range transmission as follows: Remove cap screws (22—Fig. 15) and front shield (23). Remove snap ring (1), shims (2)

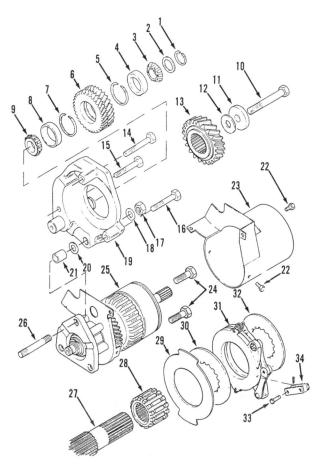

Fig. 15—Exploded view of front drive axle clutch, drive gears and components.

1. Snap ring	13. Drive gear	
2. Shim	14. Cap screw	24. Cap screws
3. Bearing cone	15. Cap screw	25. Clutch assy.
4. Bearing cup	16. Cap screw	26. Stud
5. Snap ring	17. Nut	27. Bevel pinion shaft
6. Idler gear	18. Washer	28. Brake hub
7. Snap ring	19. Park brake	29. Backing plate
8. Bearing cup	housing	30. Brake disc
9. Bearing cone	20. Shim	31. Actuator assy.
10. Cap screw	21. Dowel	32. Brake disc
11. Spacer	22. Cap screws	33. Pin
12. Shim	23. Front shield	34. Links

and bearing cone (3). Turn cap screw (10), drive gear (13) and pinion shaft (27) until idler gear (6) can be removed. See Fig. 16. Gear teeth will align in only one position for gear removal. Remove idler gear (6—Fig. 15) and bearing cone (9). Remove bearing cups (4 and 8) and snap rings (5 and 7) from idler gear. Remove cap screw (10), spacer (11), shims (12) and drive gear (13). Remove park brake pin (33). Remove park brake housing cap screws (15 and 16) and install two long alignment studs (CAS-1995), then remove cap screw (14). Remove nut (17), washer (18) and cap screws (24). Carefully slide the clutch assembly and park brake assembly forward until clutch assembly can be removed. Do not lose shims (20). If dowel (21) holds brake housing to clutch assembly (25), use extreme care when removing the units as an assembly. Lay the assembly on a bench and separate. Remove park brake components (28 through 32).

To disassemble the front-wheel drive clutch, use a sharp knife to cut and remove seal ring (1—Fig. 17). Remove snap ring (2), thrust washers (3 and 5) and thrust bearing (4). Remove snap ring (6) and shims (7), then using a suitable press, remove gear and clutch housing (19) from bearing housing (10). Remove bearing cone (8). Remove seal (17), spring (16), cap screws (15), shield (14) and cap screws (11). Remove bearing cups (9 and 12) from bearing housing (10). Using a suitable puller, remove bearing cone (13) from gear assembly (19). Remove bearings (18 and 20). Remove thrust washer (21), thrust bearing (22) and recessed thrust washer (23). Using a press (Fig. 17A), compress Belleville springs (39—Fig. 17). Remove snap ring (24) and thrust washer (25), then release press and remove shaft (40) and Belleville washers (39). Remove and discard "O" rings (36 and 38) from shaft (40). Remove snap ring (26), shims (27)

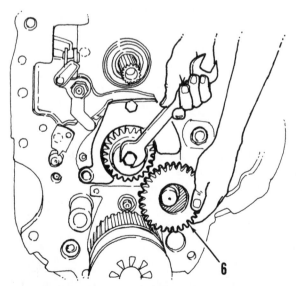

Fig. 16—Rotate drive pinion shaft until idler gear (6—Fig. 15) can be removed. Idler gear can be removed in only one position.

and backing plate (29) with "O" ring (28). Remove friction plates (30), separator plates (31) and pressure plate (32). Remove piston hub (33), "O" ring (34) and seal (35) from piston (37).

Clean and inspect all parts and renew any showing excessive wear or other damage. Check parts against the following new part dimensions:

Friction plate thickness. . . . . 2.45-2.60 mm
(0.096-0.102 in.)
Separator plate thickness . . . . 2.16-2.31 mm
(0.085-0.091 in.)
Belleville spring free height . . . . 11.0 mm
(0.433 in.)

Use all new "O" rings and seals in final assembly.

Before final assembly, determine thickness of shim pack (27—Fig. 17) to be installed. Place shaft (40) in a protective vise with splined end down. Do not install "O" rings, seals, Belleville springs or shims at this time. Install piston (37—Fig. 18) on shaft (40) until seated on shoulder. Install piston hub (33) and pressure plate (32). Starting with a separator plate (31), alternately install separator plates and friction plates (30). Install backing plate (29) and snap ring (26). Push backing plate up against snap ring and measure distance (B). Subtract 1.0-1.25 mm (0.040-0.049 in.) from measurement (B). The result will be the correct thickness of shim pack (27—Fig. 17) to be installed. Remove all installed parts from the shaft.

Install Belleville springs (39) on shaft (40) making sure that fingers on the springs are aligned. Lubricate and install new "O" rings (36 and 38) in grooves on shaft. Install piston (37) over "O" rings on shaft. Lubricate and install new "O" ring (34) and seal (35) on piston hub (33), then install piston hub and pressure plate (32). Starting with a separator plate (31), alternately install separator plates and friction plates (30). Install backing plate (29), previously determined shim pack (27) and snap ring (26). Use gear and clutch drum (19) to align friction plates (30). Use a press (Fig. 17A) to compress the Belleville springs (39—Fig. 17) and install thrust washer (25) and snap ring (24). Remove unit from press and install recessed thrust washer (23) with recess toward snap ring. Install new "O" ring (28) in groove of backing plate (29). Install thrust bearing (22) and thrust washer (21). If bearings (18 and 20) were removed, press bearings into bore of gear and clutch drum (19) until bottomed against shoulders. Install bearings cups (9 and 12) in bearing housing (10). Install cap screws (11) with washers in holes at straight edge of bearing housing. Install shield (14) and tighten cap screws (15) securely. Heat bearing cone (13) to a temperature of 150° C (300° F) and install on gear and drum assembly (19). Install spring (16) and seal (17) onto bearing housing (10). Install bearing housing on gear and drum assembly (19) and press bearing cone (8) on the gear shaft until there is no shaft end play. Install snap ring (6) and measure the clearance between bearing cone shoulder and snap ring. Subtract 0.06 mm (0.0024 in.) from the measurement. The result will be the correct thickness of shim pack (7) to be installed. Remove snap ring, install shims (7) and reinstall snap ring. Using a soft hammer, seat bearing cone (8) against shims. Using a dial indicator, check end play of bearing housing (10), which should be 0.025-0.102 mm (0.001-0.004 in.). If not, recheck shim measurement. Install bearing housing with gear and clutch drum onto clutch assembly. Install thrust washers (3 and 5) with thrust bearing (4) between the washers. Install snap ring (2) and seal ring (1).

Refer to Fig. 15 and install hollow dowel (21) without shim (20) into the bearing housing. Install

park brake housing (19) and slide dowel into brake housing. Install assembly over guide studs. Install cap screws (24) and tighten to 310-380 N·m (230-280 ft.-lbs.). Install upper park brake cap screw (14). Remove the two alignment guide studs and install cap screws (15 and 16). Tighten cap screws (14, 15 and 16) to a torque of 125-150 N·m (92-110 ft.-lbs.). Do not install nut (17) at this time. Measure the gap between the clutch bearing housing and the park brake housing above the hollow dowel. This measured gap is the correct thickness of shim pack (20) to be installed.

Install snap rings (5 and 7—Fig. 15) and bearing cups (4 and 8) into idler gear (6). Install bearing cone (9) and idler gear (6), then install bearing cone (3) and snap ring (1). Using a dial indicator, measure end play

Fig. 17—Exploded view of front-wheel drive clutch assembly.

1. Seal ring
2. Snap ring
3. Thrust washer
4. Thrust bearing
5. Thrust washer
6. Snap ring
7. Shims
8. Bearing cone
9. Bearing cup
10. Bearing housing
11. Cap screws
12. Bearing cup
13. Bearing cone
14. Rear shield
15. Cap screw
16. Spring
17. Seal
18. Bearing
19. Clutch housing & gear
20. Bearing
21. Thrust washer
22. Thrust bearing
23. Recessed thrust washer
24. Snap ring
25. Thrust washer
26. Snap ring
27. Shims
28. "O" ring
29. Backing plate
30. Friction plates
31. Separator plates
32. Pressure plate
33. Piston hub
34. "O" ring
35. Seal
36. "O" ring
37. Piston
38. "O" ring
39. Belleville springs
40. Drive shaft

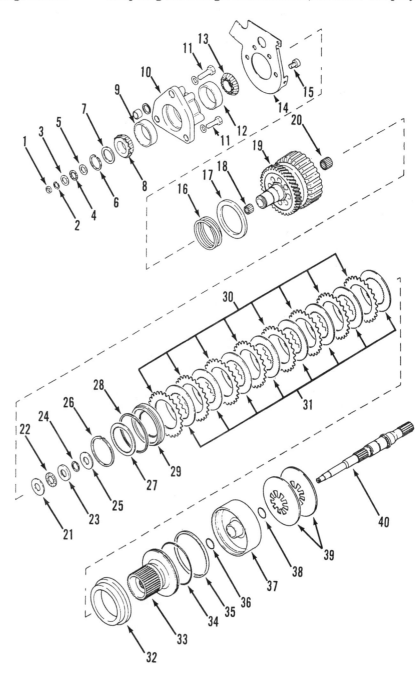

of idler gear. Subtract 0.06 mm (0.0024 in.) from this measurement to obtain correct thickness of shim pack (2) to be installed. This will provide idler gear end play of 0.025-0.102 mm (0.001-0.004 in.).

Remove snap ring (1—Fig. 15), bearing cone (3), idler gear (6) and bearing cone (9). Lay aside with the determined shim pack (2). Remove cap screws (15 and 16) and reinstall the long alignment studs. Remove upper mounting cap screw (14). Remove cap screws (24) and carefully remove the clutch and park brake housing. Separate the assemblies and install the predetermined shim pack (20) over the dowel (21), then slide assemblies back together over the dowel. Install brake hub (28) onto drive pinion shaft (27) with recessed end to the rear. Install backing plate (29), friction plate (30), actuator (31) and second friction plate (32). Reinstall park brake and clutch assembly. Install and tighten clutch mounting cap screws (24) and nut (17) with washer (18) to a torque of 310-380 N·m (230-280 ft.-lbs.). Install cap screw (14), remove alignment guide studs and install cap screws (15 and 16). Tighten cap screws to 125-150 N·m (92-110 ft.-lbs.) torque. Connect park brake linkage. Install drive gear (13) with long side of hub forward. Install pinion shaft cap screw (10), spacer (11) and shims (12) and tighten cap screw to a torque of 335-375 N·m (247-277 ft.-lbs.). Install bearing cone

(9), idler gear (6), bearing cone (3), predetermined shim pack (2) and snap ring (1). Install clutch shield (23), then install and tighten cap screws (22) securely.

Reassemble tractor by reversing disassembly procedure. Apply a coat of Loctite 515 between speed transmission and range transmission housings. Tighten speed transmission to range transmission housing bolts to a torque of 312 N·m (230 ft.-lbs.). Tighten cab mounting nuts and bolts to 310-380 N·m (230-280 ft.-lbs.) torque. Tighten rear wheel nuts to 435-515 N·m (320-350 ft.-lbs.). Fill transmission with Hy-Tran Plus fluid. Capacity is approximately 76.0 liters (80.3 U.S. qt.).

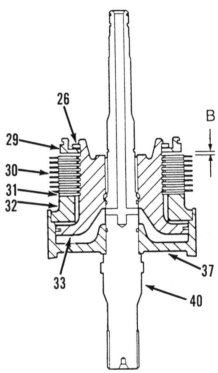

Fig. 18—View showing procedure for determining thickness of shim pack (27—Fig. 17) to be installed.

26. Snap ring	32. Pressure plate
29. Backing plate	33. Piston hub
30. Friction plates	37. Piston
31. Separator plates	40. Shaft

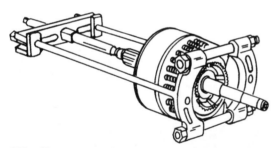

Fig. 17A—Use a screw-type press as shown to compress Belleville springs for removing or installing snap ring (24—Fig. 17).

POWER STEERING SYSTEM

All models are equipped with hydrostatic power steering using a Danfoss steering unit. The variable displacement piston-type pump furnishes pressurized oil to the compensator valve where a portion of the oil is diverted to the steering unit at a pressure of 16,893-17,582 kPa (2450-2550 psi). One double-acting steering cylinder is used on all models.

FLUID, FILTER AND BLEEDING

All Models

13. The transmission and hydraulic system fluid is used in the power steering system. A dipstick, located at the right rear of the tractor rear housing, indicates fluid level. Use Hy-Tran Plus fluid in all models. Refer to CONDENSED SERVICE DATA for reservoir capacity and to HYDRAULIC SYSTEM SECTION for additional hydraulic system information. An externally mounted, screw-on-type pressure filter is located on the right side of speed transmission housing.

The manufacturer recommends that transmission and hydraulic fluid be drained and new Hy-Tran Plus fluid be installed after each 1000 hours of operation. Filter element (14—Fig. 19) should be changed whenever service monitor on instrument cluster indicates a clogged filter or at least once each 1000 hours of operation.

To change the filter element (14), use a strap wrench and rotate element counterclockwise to remove. Lubricate gasket (13) on new element with clean oil and install element on adapter (12) until gasket contacts filter manifold. Then, hand tighten element an additional one-third turn.

Start engine and check for leaks. To bleed the steering system, operate engine at low idle speed and rotate steering wheel from lock to lock several times until steering action is firm. Check transmission oil level and add Hy-Tran Plus oil as necessary.

TROUBLESHOOTING

All Models

14. Some troubles that may occur in the operation of the power steering system and their possible causes are as follows:

1. No power steering or steers slowly.
 a. Binding mechanical linkage.
 b. Excessive load on front wheels and/or air pressure low in front tires.

c. Steering cylinder piston seal faulty or cylinder damaged.
 d. Faulty hand pump.
 e. Faulty hydraulic supply pump.
 f. Faulty priority valve in compensator valve assembly.

2. Will not steer manually.
 a. Binding mechanical linkage.
 b. Excessive load on front wheels and/or air pressure low in front tires.
 c. Pumping element in hand pump faulty.
 d. Faulty seal in steering cylinder or cylinder damaged.

3. Hard steering through complete cycle.
 a. Low pressure from supply pump.
 b. Internal or external leakage.
 c. Faulty steering cylinder.

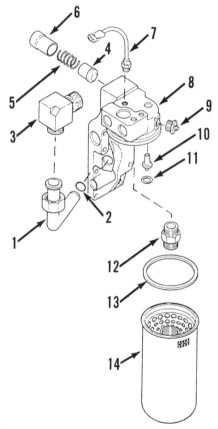

Fig. 19—Exploded view of hydraulic pressure filter and relative components.

1. By-pass tube	8. Filter manifold
2. "O" ring	9. Pressure switch
3. Elbow	10. By-pass valve
4. Lube priority piston	11. Retaining ring
5. Spring	12. Adapter
6. Sleeve	13. Gasket
7. Temperature switch	14. Filter element

d. Binding mechanical linkage.

e. Excessive load on front wheels and/or air pressure low in front tires.

4. Momentary hard or lumpy steering.

a. Air in power steering circuit.

STEERING OPERATING PRESSURE

All Models

15. To check the power steering operating pressure, refer to Fig. 20 and remove plug (5). Install a 30,000 kPa (4000 psi) pressure gauge. Operate tractor until the hydraulic oil reaches a temperature of 49° C (120° F). Operate engine at 1500 rpm, then turn and hold steering on full lock. Check pressure gauge reading, which should be 16,893-17,582 kPa (2450-2550 psi). If pressure is not correct, remove cap from relief valve (3), loosen locknut and turn adjusting screw as required to obtain correct pressure. If pressure is too low and turning adjusting screw clockwise will not increase pressure, remove test gauge and install plug (5). Refer to Fig. 21 and remove plug (P) from pump compensator (PC). Install the 30,000 kPa (4000 psi) pressure gauge. Operate engine at 1500 rpm and operate remote valve levers for maximum pressure. Pressure should be 18,600-19,300 kPa (2700-2800 psi). If not, remove cap from relief valve (4—Fig. 20), loosen locknut and turn adjusting screw to obtain correct pressure. Tighten locknut and install cap. Remove gauge and install plug. Recheck steering pressure as previously outlined and adjust as necessary.

NOTE: If the main remote/hitch pressure is too low (below steering pressure), adjusting the steering pressure relief valve will not increase steering pressure.

OPERATIONAL TESTS

All Models

The following tests are valid only if the power steering system is void of any air. If necessary, bleed system as outlined in paragraph 13 before performing tests.

16. HAND PUMP. With power steering supply pump inoperative (engine not running), attempt to steer manually in both directions.

NOTE: Manual steering with power steering supply pump inoperative, will require high steering effort.

If manual steering can be accomplished with supply pump inoperative, it can be assumed that the manual steering pump is operating satisfactorily.

17. STEERING WHEEL SLIP (CIRCUIT TEST). Steering wheel slip is the term used to describe the inability of the steering wheel to hold a given position without further steering movement.

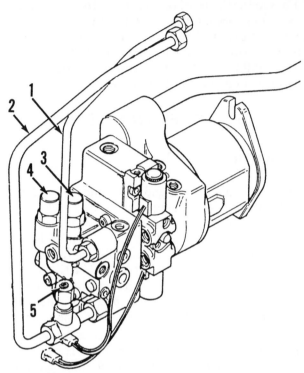

Fig. 20—To check steering pressure, remove plug (5) and install a 30,000 kPa (4000 psi) pressure gauge. Refer to text.

1. Steering pilot line
2. Steering pressure line
3. Steering pressure relief valve
4. Remote/hitch pressure relief valve
5. Plug

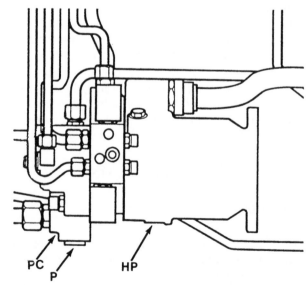

Fig. 21—To check remote/hitch maximum high-pressure or standby low-pressure, remove plug (P) from bottom of pump compensator and install test gauge.

P. Plug
PC. Pump compensator
HP. Hydraulic pump

Wheel slip is generally due to leakage, either internal or external, or a faulty hand pump or steering cylinder. Some steering wheel slip, with hydraulic fluid at operating temperature, is normal and permissible. A maximum of one revolution per minute is acceptable.

To check for steering wheel slip, proceed as follows: Operate tractor at 1500 rpm until hydraulic fluid is warmed to a temperature of 49° C (120° F). Remove steering wheel cap, then turn front wheels until they are against the stop. Attach a torque wrench to steering wheel nut. Apply 8 N·m (72 in.-lbs.) of force to steering shaft in same direction as front wheels are positioned against stop, then count number of revolutions of steering wheel in a period of one minute. Use same procedure and check steering wheel slip in opposite direction. Up to one revolution in one minute in either direction is acceptable and system can be considered as operating satisfactorily. If steering wheel revolutions per minute exceed the maximum, record total rpm for use in checking steering cylinder or hand pump as outlined in the following paragraph.

18. STEERING CYLINDER TEST. If steering wheel slip, as checked in paragraph 17, exceeds the maximum one revolution per minute, proceed as follows: Disconnect and plug lines to steering cylinder. Make sure that fluid is at operating temperature, then repeat steering wheel slip test as outlined in paragraph 17. If steering wheel slip is less than that recorded in previous test, overhaul or renew steering cylinder. If steering wheel slip, with steering cylinder disconnected, still exceeds maximum one revolution per minute, overhaul or renew steering hand pump.

HYDRAULIC SUPPLY PUMP

All Models

19. The power source for the hydrostatic steering system is the variable displacement axial piston, 78.5 L/m (20.7 U.S. gpm) pump. This pump also furnishes pressurized oil for the main hydraulic system. Refer to HYDRAULIC SYSTEM SECTION for information and service procedures for the pump.

STEERING CONTROL UNIT (HAND PUMP)

All Models

20. REMOVE AND REINSTALL. To remove the hand pump assembly, first disconnect battery cables, then drive roll pin from throttle pedal. Remove cab floor cover. Remove steering column side covers and steering tilt control lever. Remove screws and steering column center panel. Identify and disconnect the four steering tubes and the signal port tube from the

hand pump. Plug or cap all openings immediately to prevent dirt from entering the system. Remove right side steering pivot bolt. Unbolt and remove right mounting bracket. Remove roll pin from the coupling. Remove remaining two bolts and lift out hand pump assembly.

To reinstall the hand pump assembly, place pump in position and install (but do not tighten) the two left mounting bolts. Align holes in input shaft and "U" joint and install roll pin. Install steering column right mounting bracket. Install right side hand pump bolts, then tighten all four mounting bolts. Install and tighten steering column pivot bolt to a torque of 32-37 N·m (23-27 ft.-lbs.). Reconnect steering lines and

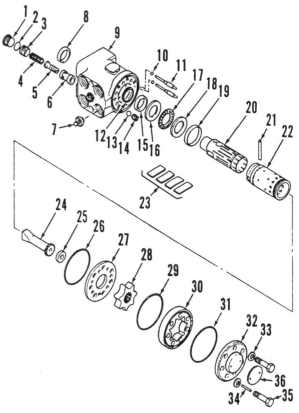

Fig. 22—Exploded view of Danfoss steering hand pump. Two shock valves (1 through 6) are used.

1.	Plug	19.	Retaining ring
2.	"O" ring	20.	Spool
3.	Spring retainer	21.	Cross-pin
4.	Spring	22.	Sleeve
5.	Shock valve	23.	Plate springs
6.	Seat	24.	Drive shaft
7.	Check valve	25.	Spacer
8.	Dust seal	26.	"O" ring
9.	Housing	27.	Valve plate
10.	Ball (2)	28.	Rotor
11.	Suction valve pin (2)	29.	"O" ring
12.	"O" ring	30.	Stator
13.	Check ball	31.	"O" ring
14.	Threaded bushing	32.	End cover
15.	Oil seal	33.	Seal washer (7)
16.	Thrust washer	34.	Roll pin
17.	Thrust bearing	35.	Special cap screw
18.	Thrust washer	36.	Name plate

signal port line. Connect battery cables and start engine. Check for leaks and bleed air from system by rotating steering wheel lock to lock several times.

> **WARNING: If steering hand pump was serviced and incorrectly reassembled, pump could operate as a motor. In this case, with engine running, steering wheel will rotate continuously. Keep hands clear of steering wheel. Immediately stop engine and correct the problem.**

The balance of installation is the reverse of removal procedures.

21. OVERHAUL. Although the output is different for the hand pumps used on two-wheel drive and four-wheel drive tractors, overhaul procedures are the same. Two-wheel drive models use a 10.0 L/m (2.6 U.S. gpm) hand pump and four-wheel drive models use a 16.0 L/m (4.2 U.S. gpm) pump. Some parts are not interchangeable.

To disassemble the removed steering hand pump, clean exterior of assembly, then unbolt and remove mounting bracket and input steering shaft. Scribe match marks across end cover (32—Fig. 22), stator (30), valve plate (27) and housing (9) for aid in correct reassembly. Clamp valve housing in a soft-jawed vise

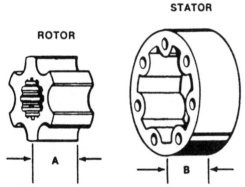

Fig. 23—When checking rotor and stator, measure distance "A" on rotor and distance "B" on stator. Refer to text.

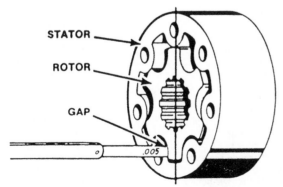

Fig. 24—Use a feeler gauge to measure gap between rotor and stator.

in inverted position and remove end cover cap screws. Remove end cover (32), stator (30), spacer (25), rotor (28), valve plate (27) and drive shaft (24). Using a screwdriver and magnet, unscrew and remove threaded bushing (14) and ball (13). Remove unit from vise and shake suction pins (11) and balls (10) from housing. DO NOT use a magnet to remove pins and balls. Push valve spool assembly from housing. Remove thrust washers (16 and 18), needle bearing (17) and retaining ring (19) from end of spool. Remove cross-pin (21) from spool, then slide spool (20) from sleeve (22). Remove inner and outer plate springs (23). Remove oil seal (15), "O" ring (12) and dust seal (8) from housing. Using an Allen wrench, remove plug (1) with "O" ring (2). Note position of spring retainer (3), then unscrew while counting the number of turns needed to remove the retainer. Record the number of turns to aid in reassembly. Remove spring (4) and shock valve (5). DO NOT remove seat (6) or check valve (7) as they are not available as replacement parts.

Clean and inspect all parts for excessive wear or other damage. If housing (9), seat (6), check valve (7), spool (20), sleeve (22), rotor (28) or stator (30) are not suitable for further service, renew complete hand pump assembly as these parts are not serviced separately. All "O" rings, seals and plate springs are available in a seal kit. Check rotor (28) and stator (30) for wear as follows: Refer to Fig. 23 and measure thickness of rotor at (A) and thickness of stator at (B). If (A) is 0.051 mm (0.002 in.) less than (B), renew hand pump. Refer to Fig. 24 and place rotor in stator as shown. Using a feeler gauge, measure gap as shown. If gap is 0.127 mm (0.005 in.) or more, renew hand pump assembly.

Reassemble by reversing the disassembly procedure keeping the following points in mind: Lubricate all internal parts with Hy-Tran Plus fluid and coat all

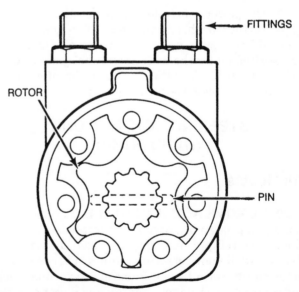

Fig. 25—View showing correct rotor to cross-pin position for correct assembly.

"O" rings with petroleum jelly during reassembly. When installing plate springs (23—Fig. 22), first install the two flat springs. Then, install the two curved springs (arches back-to-back) in between the two flat springs. When installing rotor (28) on shaft (24), make certain that cross-pin in shaft aligns with center of valleys in rotor. See Fig. 25. Align match marks and tighten end cover cap screws (35) to a torque of 30-35 N·m (22-26 ft.-lbs.).

Install input steering shaft and mounting bracket, and tighten cap screws securely.

STEERING CYLINDER

Models With Two-Wheel Drive

22. R&R AND OVERHAUL. To remove the steering cylinder, loosen jam nut (22—Fig. 26). Disconnect steering lines from cylinder and cap or plug all openings. Disconnect ball joint (23) from steering arm, then remove ball joint from steering cylinder. Remove nut (1) and lift out steering cylinder.

To disassemble the steering cylinder, remove snap ring (21), push seal gland (17) inward and remove retaining ring (20). Grasp end of cylinder rod (11) and pull outward to remove gland assembly (15 through 19) and the piston rod assembly (11 through 14). Remove wiper ring (19), seal ring (18), backup ring

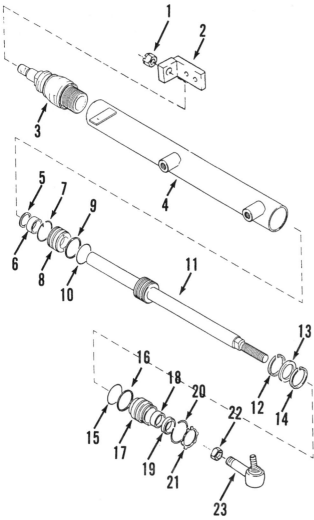

Fig. 26—Exploded view of steering cylinder used on two-wheel-drive tractors.

1.	Nut	13.	Seal ring
2.	Bracket	14.	Wear ring
3.	Cylinder end	15.	"O" ring
4.	Cylinder tube	16.	Backup ring
5.	Backup ring	17.	Seal gland
6.	Seal ring	18.	Seal ring
7.	Retaining wire	19.	Wiper ring
8.	Seal gland	20.	Retaining ring
9.	Backup ring	21.	Snap ring
10.	"O" ring	22.	Jam nut
11.	Piston rod	23.	Ball joint
12.	Wear ring		

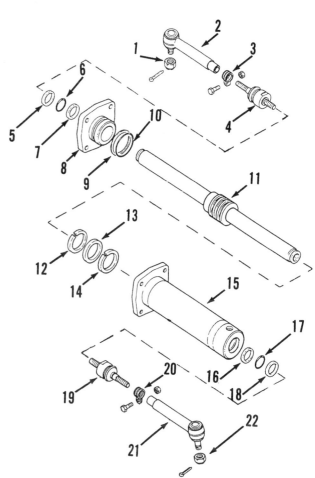

Fig. 27—Exploded view of steering cylinder used on tractors equipped with front drive axle.

1.	Nut	12.	Wear ring
2.	Tie rod end (L.H.)	13.	Piston seal ring
3.	Clamp	14.	Wear ring
4.	Tie rod stud (L.H.)	15.	Cylinder tube
5.	Wiper seal	16.	Rod seal
6.	Backup washer	17.	Backup ring
7.	Rod seal	18.	Wiper seal
8.	End cap	19.	Tie rod stud (R.H.)
9.	Backup ring	20.	Clamp
10.	"O" ring	21.	Tie rod end (R.H.)
11.	Piston rod	22.	Nut

(16) and "O" ring (15) from gland (17). Remove seal ring (13) and wear rings (12 and 14) from piston rod. Remove cylinder end (3) from cylinder tube (4). Using special tool (CAS-3461) and a long extension inside the cylinder tube, rotate seal gland (8) until retaining wire (7) is removed through slot in cylinder tube. Remove the gland assembly. Remove backup ring (5), seal ring (6), backup ring (9) and "O" ring (10) from gland (8).

Clean and inspect seal glands (8 and 17), piston rod (11) and cylinder tube (4) for scoring or other damage and renew as necessary. A seal kit including all "O" rings, seals and retaining rings is available.

When reassembling, lubricate all internal parts with clean Hy-Tran Plus oil and reverse the disassembly procedure. Apply Loctite R.T.V. sealant 595 to the seal gland retaining wire slot. Apply Loctite 270 to the threads of cylinder end (3) and tighten to a torque of 400 N•m (295 ft.-lbs.).

When reinstalling cylinder assembly, tighten nut (1) to a torque of 380-420 N•m (280-310 ft.-lbs.). Tighten jam nut (22) to 200 N•m (148 ft.-lbs.). Connect ball joint (23) to steering arm and tighten to 400 N•m (295 ft.-lbs.). Reconnect steering lines and bleed air from steering system.

Models With Front Drive Axle

23. R&R AND OVERHAUL. To remove the steering cylinder, first raise front of tractor, support with two axle stands and remove front wheels. Remove nuts (1 and 22—Fig. 27) and disconnect tie rod ends (2 and 21) from steering arms. Remove tie rod studs (4 and 19) from ends of piston rod (11). Disconnect steering lines at cylinder. Remove elbow fitting from right end of cylinder. Unbolt and remove cylinder out left side of drive pinion housing.

To disassemble the steering cylinder, remove end cap (8) and piston rod (11). Remove wiper seal (5), backup ring (6), rod seal (7), backup ring (9) and "O" ring (10) from end cap (8), and piston seal ring (13) and wear rings (12 and 14) from piston rod (11). Remove wiper seal (18), backup ring (17) and rod seal (16) from right end of cylinder tube (15).

Clean and inspect end cap (8), piston rod (11) and cylinder tube (15) for excessive wear or scoring and renew as necessary. A seal kit, consisting of "O" rings, wear rings seal rings, backup rings and wiper seals, is available. When reassembling, lubricate all internal parts with clean Hy-Tran Plus oil.

Reinstall cylinder by reversing the removal procedures. Tighten steering cylinder mounting bolts to a torque of 90 N•m (66 ft.-lbs.). Tighten tie rod studs (4 and 19) into ends of piston rod (11) to a torque of 300 N•m (221 ft.-lbs.). Connect tie rod ends (2 and 21) to steering arms and tighten nuts (1 and 22) to 230 N•m (170 ft.-lbs.). Use new "O" rings and connect steering lines. Install front wheels and tighten nuts to 300 N•m (221 ft.-lbs.) torque. Adjust toe-in as outlined in paragraph 10. Bleed air from steering system by operating engine at low idle speed and turn steering wheel lock-to-lock several times.

ENGINE

Model 5120 tractors prior to P.I.N. JJF1005939 are equipped with 4TA-390, 4-cylinder diesel engines having bores of 102 mm (4.02 in.), strokes of 120 mm (4.72 in.) and displacements of 3.92 L (239 cu.in.). Engines are equipped with turbochargers and aftercoolers.

Model 5120 tractors after P.I.N. JJF1005939 are equipped with 4T-390, 4-cylinder diesel engines having bores of 102 mm (4.02 in.), strokes of 120 mm (4.72 in.) and displacements of 3.92 L (239 cu. in.). Engines are equipped with turbochargers.

Model 5130 tractors are equipped with 6-590, 6-cylinder diesel engines having bores of 102 mm (4.02 in.), strokes of 120 mm (4.72 in.) and displacements of 5.9 L (359 cu. in.). Engines are naturally aspirated.

Model 5140 tractors are equipped with 6T-590, 6-cylinder diesel engines having bores of 102 mm (4.02 in.), strokes of 120 mm (4.72 in.) and displacements of 5.9 L (359 cu. in.). Engines are equipped with turbochargers.

R&R ENGINE ASSEMBLY

All Models

24. To remove engine assembly, first split tractor between engine and speed transmission as follows: Apply park brake and securely block rear wheels. Place wood wedges between front axle and front support to prevent tipping. Disconnect battery cables and remove batteries and battery box. Unbolt and remove left step assembly. Remove left rear wheel assembly. Drain fuel tank and disconnect breather hose, return hose and supply hose. Unbolt and remove fuel tank. If so equipped, remove front drive shaft and shield as outlined in paragraph 9. Remove plugs and drain oil from transmission and flywheel housings. Raise the hood and remove exhaust extension pipe and rear hood cover. Disconnect hood release cable and move cable out of the way. Disconnect right steering tube and the supply and return oil cooler tubes. Disconnect wires and cable from starter motor. Drain cooling system, then shut off and disconnect heater hoses. Disconnect throttle cable from injection pump. Disconnect left steering tube. Loosen clamp and remove brake return hose from housing adapter. Disconnect engine wiring harness connector from fire wall. Disconnect ground cable and relay cables. If so equipped, disconnect air conditioning hoses at quick couplers. Loosen rear cab or platform mounting nuts and remove front cab or platform mounting bolts. Carefully tilt cab or platform and install blocks between cab or platform and mounting frame. If so equipped, remove tractor front end weights. Attach front split stand (CAS 10852) to

support front of tractor and block up under front of speed transmission housing. Remove the four upper transmission-to-engine mounting bolts and the left and right outer side-rail bolts. Remove remaining transmission-to-engine bolts and carefully separate engine from transmission.

Remove exhaust muffler assembly and rear hood support bracket. Disconnect air restriction switch wire and remove air inlet tube. Disconnect air conditioning compressor clutch wire and high-pressure switch wire. Unbolt and remove compressor and carefully place compressor in a safe horizontal position. Disconnect injection pump fuel shut-off wire, oil pressure switch wire and coolant temperature switch wire. Remove upper and lower radiator hoses. Disconnect wires from alternator terminals D+ and W, and move wiring harness forward out of the way. Attach a sling to engine lifting eyes and connect to an overhead hoist. Unbolt and lift engine from frame.

Reinstall engine by reversing removal procedure. Clean transmission and engine mounting flanges, and apply a bead of Loctite 515 to engine mounting flange. Use two (CAS-1995) alignment studs to guide engine and transmission together. Tighten 16 mm bolts to a torque of 335-375 N·m (247-277 ft.-lbs.) and 12 mm bolts to a torque of 135-150 N·m (100-110 ft.-lbs.). Tighten cab mounting nuts and bolts to a torque of 310-380 N·m (230-280 ft.-lbs.). Tighten rear wheel nuts to 435-515 N·m (320-350 ft.-lbs.). Fill transmission with Hy-Tran Plus fluid. Capacity is approximately 76.0 liters (80.3 U.S. qt.). The balance of assembly is the reverse of disassembly procedure.

CYLINDER HEAD

All Models

25. REMOVE AND REINSTALL. To remove the cylinder head, first disconnect battery cables. Then, raise hood and remove side shields, exhaust extension pipe and rear hood cover. Drain cooling system. Unbolt and remove muffler, disconnect air restriction switch wire and remove intake air tube. Remove compressor belt, if so equipped, then lift belt tensioner and remove fan belt. Unbolt and remove fan, compressor drive pulley and fan pulley. Unbolt and remove fan bracket, belt tensioner and belt tensioner bracket. Remove alternator and mounting bracket. Remove upper radiator hose, thermostat housing with thermostat and front lifting bracket. Disconnect oil supply and drain tubes from turbocharger, if so equipped, then unbolt and remove turbocharger assembly. Unbolt and remove exhaust manifold. Disconnect and remove fuel injection lines and fuel leak-off lines. Cap or plug all openings. Remove fuel injectors from cylinder head. Remove inlet and outlet lines for fuel filters and cover all openings. Remove fuel filters, then remove retaining nut and lift off filter housing. On Model 5120 prior to P.I.N. JJF1005939, unbolt and remove aftercooler and, on all other models, unbolt and remove intake manifold cover. Then, on all models, remove valve covers. Remove rocker arm bolts, then remove rocker arm assemblies and push rods. Unbolt and lift off cylinder head.

Clean engine block and cylinder head mating surfaces, removing carbon or other deposits. Check cylinder head for cracks or other signs of damage in the area of fire ring contact. Using a straightedge and feeler gauge, check cylinder head at each side and between cylinders for flatness. Resurface cylinder head if it is warped more than 0.010 mm (0.0004 in.) in any 50 mm (2.0 in.) diameter area or if it is distorted more than 0.075 mm (0.003 in.) overall, end-to-end or side-to-side. Install new cylinder head if thickness is less than 93.75 mm (3.691 inches) after machining.

Check cylinder block surface for flatness and resurface as outlined in paragraph 40 if not flat within 0.075 mm (0.003 inch).

Cylinder head gaskets of three different thicknesses are available to compensate for cylinder block height. Make sure correct gasket is installed. Measure amount of piston stand-out above cylinder block at top dead center of several cylinders. If piston stand-out is less than 0.660 mm (0.026 inch), standard thickness head gasket, marked with one notch (Fig. 28), should be installed. Cylinder head gasket marked with two notches should be installed if piston stand-out is 0.661-0.910 mm (0.0261-0.0358 inch) and head gasket with three notches should be installed if stand-out is 0.911-1.166 mm (0.0359-0.0457 inch). Cylinder block should not be lowered far enough to cause piston protrusion of more than 1.166 mm (0.0457 inch). Machining cylinder head does not affect selection of cylinder head gasket.

Install correct head gasket and cylinder head, then lubricate bolt threads and install head bolts, except for rocker arm head bolts. Install push rods and rocker arm assemblies. Lubricate threads and install rocker arm head bolts and rocker arm bracket bolts.

On Model 5120 tractors, tighten head bolts following the sequence shown in Fig. 29. Tighten head bolts in three steps as follows: Torque to 40 N·m (30 ft.-lbs.) in step one, 80 N·m (60 ft.-lbs.) in step two and 129 N·m (95 ft.-lbs.) in step three. Tighten rocker arm bracket bolts to a torque of 24 N·m (18 ft.-lbs.).

On Models 5130 and 5140, tighten head bolts following the sequence shown in Fig. 30. Tighten head bolts in four steps as follows: Tighten all head bolts to 41 N·m (30 ft.-lbs.) in step one and to 90 N·m (66 ft.-lbs.) in step two. In step three, tighten all head bolts except numbers 4, 5, 12, 13, 20 and 21 to a torque of 120 N·m (88 ft.-lbs.). In step four, tighten all head bolts an additional 90°. Tighten rocker arm bracket bolts to a torque of 24 N·m (18 ft.-lbs.).

On all models, using a new gasket, install intake cover or aftercooler if so equipped. Refer to Fig. 31 or Fig. 32 and tighten retaining bolts to a torque of 24 N·m (18 ft.-lbs.). Refer to paragraph 66 for procedure on installation of fuel injectors and to paragraph 73 for installation of turbocharger. Adjust valve tappet gap as outlined in paragraph 27.

Complete the assembly by reversing the disassembly procedure. Other tightening torques are as follows:

Filter housing nut . . . . 32 N·m
(24 ft.-lbs.)

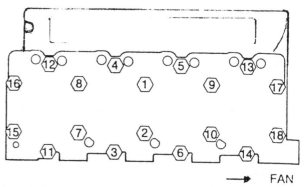

Fig. 29—Cylinder head bolt tightening sequence for Model 5120 tractors.

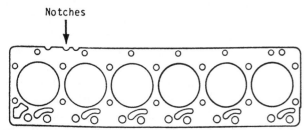

Fig. 28—View showing cylinder head gasket thickness identification notches. Six-cylinder gasket is shown; four-cylinder gasket is similar.

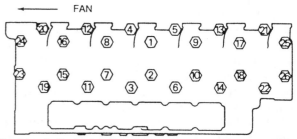

Fig. 30—Cylinder head bolt tightening sequence for Model 5130 and 5140 tractors.

Thermostat housing bolts . . . . 24 N·m
(18 ft.-lbs.)
Exhaust manifold bolts . . . . 43 N·m
(32 ft.-lbs.)
Belt tensioner bracket bolts . . . . 24 N·m
(18 ft.-lbs.)
Belt tensioner bolt . . . . 43 N·m
(32 ft.-lbs.)
Fan pulley bracket bolts . . . . 24 N·m
(18 ft.-lbs.)
Fan retaining bolts . . . . 43 N·m
(32 ft.-lbs.)
Valve cover bolts . . . . 24 N·m
(18 ft.-lbs.)

Refer to Fig. 33 or Fig. 34 when installing exhaust manifold.

VALVES AND SEATS

All Models

26. Intake and exhaust valves seat directly in the cylinder head. However, service seat inserts are available for both intake and exhaust valves. Stem seals are used on all valves. Check valves against the following specifications:

INTAKE
Face angle . . . . 29°
Seat angle . . . . 30°
Stem diameter . . . . 7.960-7.980 mm
(0.313-0.314 in.)
Stem-to-guide clearance . . . . 0.039-0.079 mm
(0.0015-0.0030 in.)
Seat width . . . . 1.32-1.92 mm
(0.052-0.076 in.)

Valve tappet gap (cold) . . . . 0.25 mm
(0.010 in.)
Valve recession from face of
cylinder head (max.) . . . . 1.52 mm
(0.060 in.)

EXHAUST
Face angle . . . . 44°
Seat angle . . . . 45°
Stem diameter . . . . 7.960-7.980 mm
(0.313-0.314 in.)
Stem-to-guide clearance . . . . 0.039-0.079 mm
(0.0015-0.0030 in.)
Seat width . . . . 1.47-2.07 mm
(0.058-0.081 in.)
Valve tappet gap (cold) . . . . 0.50 mm
(0.020 in.)
Valve recession from face of
cylinder head (max.) . . . . 1.52 mm
(0.060 in.)

27. VALVE TAPPET GAP ADJUSTMENT. The two-position method of tappet gap adjustment is recommended. DO NOT attempt to adjust valves with engine operating. Clearance should be adjusted with engine cold.

On all models, use flywheel rotating tool (CAS-1690) through opening in flywheel housing and turn flywheel in normal direction of rotation while pushing inward on the lockpin. The timing lockpin is located in timing gear housing below the fuel injection pump. When lockpin moves forward into hole in camshaft gear, No. 1 piston will be at TDC on compression stroke. Both push rods on No. 1 cylinder should be loose.

On Model 5120, use a feeler gauge to measure the clearance between the rocker arm and valve stem of the valves indicated in Fig. 35A. Specified clearance is 0.25 mm (0.010 in.) for intake valves and 0.50 mm (0.020 in.) for exhaust valves. Turn rocker arm adjust-

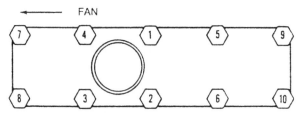

Fig. 31—Bolt tightening sequence for intake cover on aftercooler on Model 5120.

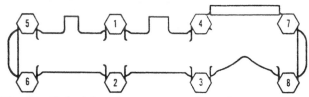

Fig. 33—Bolt tightening sequence for exhaust manifold on Model 5120.

BOLT TORQUE SEQUENCE

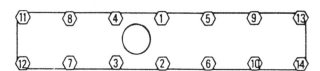

Fig. 32—Bolt tightening sequence for intake cover on Models 5130 and 5140.

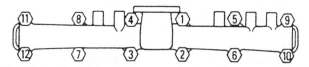

Fig. 34—Bolt tightening sequence for exhaust manifold on Models 5130 and 5140.

ing screw to adjust the clearance. Disengage timing lockpin and rotate flywheel one complete revolution. This will place No. 4 piston at TDC of compression stroke. Adjust the remaining four valves indicated in Fig. 35B.

On Models 5130 and 5140, with No. 1 piston at TDC of compression stroke, use a feeler gauge to measure

the clearance between the rocker arm and valve stem of the valves indicated in Fig. 36A. Specified clearance is 0.25 mm (0.010 in.) for intake valves and 0.50 mm (0.020 in.) for exhaust valves. Turn rocker arm adjusting screw to adjust the clearance. Disengage timing lockpin and rotate flywheel one complete revolution. This will place No. 6 piston at TDC of compression stroke. Adjust the remaining six valves indicated in Fig. 36B.

VALVE GUIDES AND SPRINGS

All Models

28. Valve guides are cast directly in cylinder head. If inside diameter of guide is worn beyond the limit of 8.089 mm (0.3185 in.), the guide bore may be machined for installation of service guides. Machine the guide bore diameter to 11.112-11.138 mm (0.4375-0.4385 in.) for the 11.150-11.163 mm (0.4389-0.4395 in.) oversize OD guide or to 13.987-14.013 mm (0.551-0.552 in.) for the 14.026-14.038 mm (0.552-0.553 in.) oversize OD guide. New guide will be an interference fit of 0.013-0.051 mm (0.0005-0.002 in.) and must be pressed into cylinder head. Press guide in from top until top end of guide is 11.25-11.75 mm (0.443-0.463 in.) above valve guide boss on cylinder head. Ream installed service guide inside diameter to 8.019-8.039 mm (0.3155-0.3165 in.). Stem seals are used on all valves.

Intake and exhaust valve springs are identical. Renew any spring that is cracked, discolored or does not meet the following specifications:

Free length	55.63 mm
	(2.19 in.)
Test length	49.25 mm
	(1.94 in.)
Test load	285-321 N
	(64-72 lbs.)

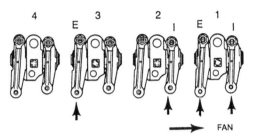

Fig. 35A—On Model 5120, with No. 1 piston at TDC of compression stroke, adjust valves indicated.

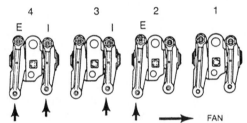

Fig. 35B—On Model 5120, with No. 4 piston at TDC of compression stroke, adjust valves indicated.

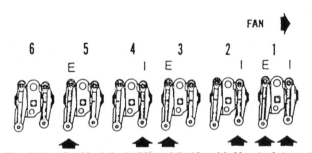

Fig. 36A—On Models 5130 and 5140, with No. 1 piston at TDC of compression stroke, adjust valves indicated.

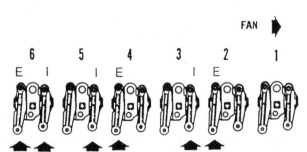

Fig. 36B—On Models 5130 and 5140, with No. 6 piston at TDC of compression stroke, adjust valves indicated.

ROCKER ARMS

All Models

29. Refer to Fig. 37 for exploded view of rocker arm and shaft assembly used on each cylinder. To disassemble the removed unit, remove snap rings, thrust washers and rocker arms. Measure rocker shaft and if worn to a diameter of less than 18.938 mm (0.745 in.), renew shaft. Measure inside diameter of rocker arm bore. If rocker arm bore is more than 19.05 mm (0.750 in.), renew rocker arm.

Tighten rocker arm head bolt to a torque of 129 N·m (95 ft.-lbs.) on Model 5120 tractors and to 120 N·m (88 ft.-lbs.) plus 90° on Model 5130 or 5140 tractors. Tighten rocker arm bracket bolt to a torque of 24 N·m

(18 ft.-lbs.) on all models. Adjust valve tappet gap as outlined in paragraph 27. Tighten rocker cover bolts to 24 N•m (18 ft.-lbs.).

CAM FOLLOWERS

All Models

30. The mushroom-type cam followers can be removed from below after first removing camshaft, oil pan and engine balancer if so equipped, as outlined in paragraphs 36, 51 and 48.

Measure outside diameter of cam follower stem. Renew cam followers if outside diameter of stems is less than 15.960 mm (0.628 in.). Measure inside diameter of cam follower bores in cylinder block. Maxi-

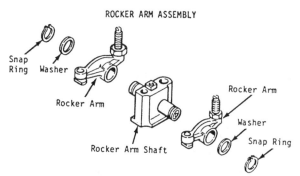

Fig. 37—Exploded view of rocker arm assembly used on each cylinder.

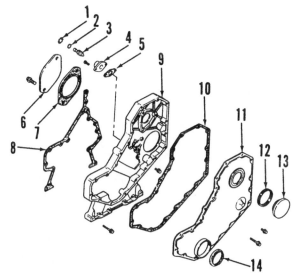

Fig. 38—Exploded view of timing gear housing, cover and related parts.

1. Retaining ring	8. Gasket
2. "O" ring	9. Timing gear housing
3. Timing lockpin	10. Gasket
4. Lockpin housing	11. Timing gear cover
5. Gasket	12. Seal ring
6. Access plate	13. Access plug
7. Gasket	14. Crankshaft seal

mum bore diameter is 16.055 mm (0.632 in.). If bores exceed the limit, renew cylinder block.

TIMING GEAR COVER

All Models

31. To remove the timing gear cover (11—Fig. 38), disconnect battery cables and remove hood, grille and side panels. If so equipped, unbolt air conditioning condenser and receiver-drier and lay rearward on engine. It is not necessary to disconnect the air conditioner lines. Disconnect air filter restriction switch and remove air intake tube. Identify and disconnect oil cooler hoses. Remove front weights, if so equipped. Drain coolant and remove upper and lower radiator hoses. Disconnect steering cylinder lines and, if so equipped, unbolt and remove front-wheel drive shaft and shield. Attach split stand to side rails and attach a hoist to front support. Unbolt and separate front axle and front support from tractor.

If so equipped, loosen compressor bolts and remove drive belt. Raise fan belt tensioner and remove fan belt. Unbolt and remove fan, compressor drive pulley and fan pulley. Unbolt and remove crankshaft pulley. Then, unbolt and remove timing gear cover.

NOTE: Refer to paragraph 45 for front crankshaft oil seal installation and timing gear cover alignment.

When reassembling, use new cover gasket and tighten cover retaining cap screws to a torque of 24 N•m (18 ft.-lbs.). Tighten crankshaft pulley bolts to a torque of 130-144 N•m (96-106 ft.-lbs.). Tighten fan retaining bolts to 43 N•m (32 ft.-lbs.). Tighten side rails to front support bolts to 134-151 N•m (100-110 ft.-lbs.) and tighten engine to frame bolts to a torque of 335-375 N•m (247-277 ft.-lbs.).

TIMING GEARS

All Models

The timing gear train consists of the crankshaft gear, camshaft gear and injection pump drive gear. Engine oil pump is mounted on front of engine block inside the timing gear housing and drives through an idler gear from crankshaft gear, becoming part of the gear train. Refer to the appropriate following paragraphs for timing, inspection and overhaul information.

32. TIMING MARKS. When installing camshaft gear and crankshaft gear, align single punch mark on crankshaft gear with double punch mark on camshaft gear (Fig. 39). When installing injection pump drive gear, align the correct letter on injection pump gear (Fig. 40) with the single punch mark on camshaft gear. Use letter "B" for Model 5120 tractors, letter "D" for Model 5130 tractors and letter "C" for Model 5140 tractors. Idler gear and oil pump gear are not marked.

When the timing pin on left rear of timing gear housing (below the injection pump) is inserted into hole in camshaft gear, No. 1 piston is at TDC on compression stroke. Crankshaft pulley is not marked.

33. CAMSHAFT GEAR. The camshaft gear (7—Fig. 41) is a shrink fit on the camshaft (2). To remove the camshaft gear, remove camshaft from engine as outlined in paragraph 36. With timing gear cover removed and before removing camshaft, use a dial indicator and check backlash between camshaft gear and crankshaft gear. Normal backlash is 0.08-0.33 mm (0.003-0.012 in.). Maximum backlash is 0.45 mm (0.018 in.). Excessive backlash could be caused by gear wear, shaft wear or camshaft bushing wear, or by a combination of factors.

To remove camshaft gear after removing camshaft, press the gear off the shaft. Refer to paragraph 37 and

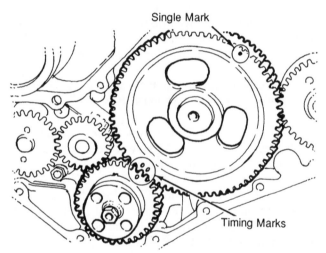

Fig. 39—To correctly "time" gear train, align single punch mark on crankshaft gear with double punch marks on camshaft gear.

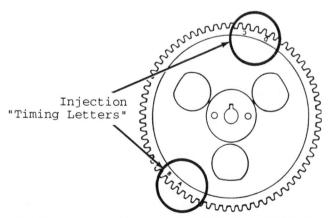

Fig. 40—Injection pump drive gear is marked with "timing letters." Refer to text for timing instructions.

check camshaft bushing and camshaft for excessive wear.

When installing camshaft gear on camshaft, heat gear approximately 45 minutes at a temperature of 120° C (250° F) in a bearing oven. Install gear with timing marks to the front and timing pin hole toward the rear.

Reinstall camshaft as outlined in paragraph 36.

34. INJECTION PUMP GEAR. To remove the injection pump drive gear, first remove the timing gear cover as outlined in paragraph 31. Using a dial indicator, check backlash between injection pump drive gear and camshaft gear. Normal backlash is 0.08-0.33 mm (0.003-0.013 in.). Maximum backlash is 0.45 mm (0.018 in.). Loosen the nut on injection pump shaft.

Using a flywheel rotating tool (CAS-1690) through hole in flywheel housing, turn flywheel in normal direction of rotation while pushing inward on the timing pin. When timing lockpin is engaged in hole in camshaft gear, loosen the injection pump lock bolt at the front left side of pump and remove the slotted tab. Then, tighten lock bolt to a torque of 30 N•m (22 ft.-lbs.). Remove nut and lockwasher from pump shaft. Using a suitable puller, such as CAS-1691A, remove pump gear.

When installing pump gear, make certain that correct timing letter (Fig. 40) on pump gear is aligned with the single timing mark on camshaft gear (Fig. 39). Use letter "B" for Model 5120 tractors, letter "D" for 5130 tractors and letter "C" for 5140 tractors.

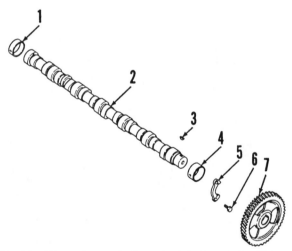

Fig. 41—View of camshaft and related parts for six-cylinder engines; camshaft for four-cylinder engines is similar. Engines are equipped with renewable front bushing (4). Service bushings (1) are available for all other journal bores.

1. Service bushings
2. Camshaft
3. Key
4. Front bushing
5. Thrust plate
6. Cap screws (2)
7. Camshaft gear

Install nut and lockwasher on pump shaft and tighten finger-tight. Loosen lock bolt on pump, insert slotted tab, then tighten lock bolt to a torque of 30 N·m (22 ft.-lbs.). Disengage timing lockpin from camshaft gear. Then, tighten pump drive gear nut to a torque of 65 N·m (48 ft.-lbs.). Refer to paragraph 31 and reinstall timing gear cover.

35. CRANKSHAFT GEAR. To remove the crankshaft gear, first remove timing gear cover. Using a dial indicator, check backlash between crankshaft gear and camshaft gear. Normal backlash is 0.08-0.33 mm (0.003-0.013 in.). Maximum allowable backlash is 0.45 mm (0.018 in.). Then, remove camshaft as in paragraph 36 and injection pump as in paragraph 62. Unbolt and remove timing gear housing (9—Fig. 38) and oil pump assembly. Using a suitable puller, remove crankshaft gear.

When reassembling, use new crankshaft gear timing pin and heat gear for approximately 25 minutes at a temperature of 120° C (250° F) in a bearing oven. Install heated gear on crankshaft. Align timing marks as outlined in paragraph 32. Use new gasket (8—Fig. 38) and install timing gear housing (9). Tighten housing retaining bolts to a torque of 24 N·m (18 ft.-lbs.). The balance of reassembly is the reverse of disassembly.

CAMSHAFT AND BEARINGS

All Models

The camshaft (2—Fig. 41) is carried in one (front) renewable bushing (4) and four bores (4-cylinder) or six bores (6-cylinder) directly in engine block. However, if the bores in block show excessive wear, engine must be completely disassembled and the block line bored for installation of service bushings.

36. R&R CAMSHAFT. To remove the camshaft and gear assembly, first remove timing gear cover as outlined in paragraph 31. Remove injection pump drive gear as outlined in paragraph 34. Unbolt and remove rocker arm covers, rocker arm assemblies and push rods. Remove cam follower cover from left side of engine, pull cam followers upward and hold them in this position by clipping clothes pins on the stems. Unbolt and remove fuel lift pump. Working through the holes in camshaft gear, unbolt and remove the camshaft thrust plate (5—Fig. 41). Withdraw camshaft and gear assembly.

Remove camshaft gear as outlined in paragraph 33 if renewal is indicated. Refer to paragraph 37 for specification data on camshaft and bearings.

When reinstalling camshaft, align timing marks as in paragraph 32. Then, using a dial indicator, check camshaft end play. End play is controlled by the thrust plate (5—Fig. 41). Camshaft end play should be 0.13-0.47 mm (0.005-0.018 in.). If end play is more than 0.47 mm (0.018 in.), renew thrust plate. Tighten camshaft thrust plate bolts (6) to a torque of 24 N·m (18 ft.-lbs.) Complete reassembly by reversing the disassembly procedure. Adjust valve clearance as outlined in paragraph 27.

37. CAMSHAFT BEARINGS. Desired camshaft clearance is 0.076-0.15 mm (0.003-0.006 in.). Measure diameter of all camshaft journals, which should be 53.987-54.013 mm (2.125-2.126 in.). If journals are worn to a diameter of 53.96 mm (2.124 in.), renew camshaft.

When checking front bushing and other journal bores in block, remove oil pan, oil pick-up tube and engine balancer, if so equipped, as outlined in paragraphs 48 through 51. Then, measure inside diameter of front bushing and other journal bores, which should be 54.107-54.133 mm (2.130-2.131 in.). If front bushing is worn to an inside diameter of 54.146 mm (2.132 in.), renew front bushing. If other journal bores in block are worn to an inside diameter of 54.146 mm (2.132 in.), block must be line bored and service bushings installed.

CONNECTING ROD AND PISTON UNITS

All Models

38. Connecting rod and piston units are removed from above after removing cylinder head, oil pan, oil pick-up tube, engine balancer, if so equipped, and connecting rod caps. Before removing, place a cylinder identification at top front of each piston. If not marked, place cylinder identification mark on connecting rod and cap on side opposite the camshaft. Remove carbon or wear ring from top of cylinders, then remove the units.

Reinstall by reversing the removal procedure. Lubricate threads and tighten connecting rod bolts on Model 5120 tractors first to a torque of 50 N·m (36 ft.-lbs.) and then to 98 N·m (72 ft.-lbs.). On Models 5130 and 5140, tighten connecting rod bolts first to a torque of 60 N·m (44 ft.-lbs.). Then, scribe six marks at equal distance (60°) around the outer diameter of the socket being used to tighten the bolts. Scribe one line on the rod cap adjacent to one of the scribe marks on the socket. Tighten the rod bolts an additional 60° (until next scribe mark on socket aligns with scribe mark on rod cap).

PISTONS AND RINGS

All Models

39. Engines are equipped with pistons having two compression rings and one oil control ring (Fig. 42) and operate directly in the cylinder block bores. Piston and ring sets are available in standard size and oversizes of 0.5 mm (0.020 in.) and 1.0 mm (0.040 in.).

NOTE: If cylinder bores are worn beyond the oversize limits, refer to paragraph 40 for information on boring and installing service sleeves.

Check the pistons, pins, rings and cylinder bores against the following specifications:

Piston type . . . . Cam-ground
Piston material . . . . Aluminum alloy
Piston skirt diameter 12 mm (0.5 in.)
 from bottom, 90° from pin—
 Standard size piston . . . . . . 101.873-101.887 mm
   (4.010-4.011 in.)
  Minimum limit . . . . 101.823 mm
   (4.008 in.)

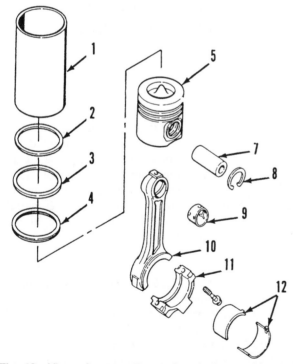

Fig. 42—View of connecting rod and piston with related parts.

1. Service sleeves
2. Top compression ring
3. Second compression ring
4. Oil control ring
5. Piston
7. Piston pin
8. Retaining ring
9. Bushing
10. Connecting rod
11. Rod cap
12. Connecting rod bearing

0.5 mm (0.020 in.) Oversize
 piston . . . . 102.373-102.387 mm
   (4.030-4.031 in.)
  Minimum limit . . . . 102.323 mm
   (4.028 in.)
1.0 mm (0.040 in.) Oversize
 piston . . . . 102.873-102.887 mm
   (4.050-4.051 in.)
  Minimum limit . . . . 102.823 mm
   (4.048 in.)
Standard cylinder bore . . . . 102.00-102.04 mm
   (4.016-4.017 in.)
 Maximum limit . . . . 102.116 mm
   (4.020 in.)
Cylinder out-of-round (max.). . . . . 0.038 mm
   (0.0015 in.)
Cylinder taper (max.) . . . . 0.076 mm
   (0.003 in.)
Piston pin diameter . . . . 39.997-40.003 mm
   (1.5747-1.5749 in.)
 Clearance in piston . . . . 0.003-0.015 mm
   (0.0001-0.0006 in.)
  Maximum limit. . . . . 0.035 mm
   (0.0014 in.)
 Clearance in rod . . . . 0.051-0.076 mm
   (0.002-0.003 in.)
  Maximum limit. . . . . 0.10 mm
   (0.004 in.)
Top compression ring—
 Type . . . . Keystone
   barrel face
 End gap. . . . . 0.40-0.70 mm
   (0.016-0.028 in.)
  Maximum limit. . . . . 0.806 mm
   (0.032 in.)
 Side clearance. . . . . Not measurable
Second compression ring—
 Type . . . . Rectangular taper face
 End gap. . . . . 0.25-0.55 mm
   (0.010-0.022 in.)
  Maximum limit. . . . . 0.806 mm
   (0.032 in.)
 Side clearance. . . . . 0.075-0.120 mm
   (0.003-0.005 in.)
  Maximum limit. . . . . 0.15 mm
   (0.006 in.)
Oil control ring—
 Type . . . . Two-piece
 End gap. . . . . 0.25-0.55 mm
   (0.010-0.022 in.)
  Maximum limit. . . . . 0.806 mm
   (0.032 in.)
 Side clearance. . . . . 0.065-0.110 mm
   (0.0025-0.0043 in.)
  Maximum limit. . . . . 0.13 mm
   (0.005 in.)

CYLINDER BLOCK AND SLEEVES

All Models

40. Cylinder block is not originally fitted with sleeves, but if cylinder bores are worn beyond the oversize limits, cylinders can be bored and service sleeves (1—Fig. 42) installed.

To install service sleeves, completely disassemble and clean cylinder block. Service sleeves do not extend to bottom of cylinder bore. Mark each cylinder bore 6.25 mm (0.250 inch) from bottom, then machine the cylinder bore to the depth of this mark. **Do not bore cylinder lower than 6.25 mm (0.250 inch) from bottom of cylinder.** Finished diameter should be 104.485-104.515 mm (4.1136-4.1148 inches).

Freeze cylinder sleeves in dry ice. Install each sleeve in cylinder bore until sleeve contacts the unmachined ridge at bottom of bore. If the installed sleeves protrude above cylinder block, machine top of sleeves until flush with top of block. Hone inside diameter of installed sleeves to 102.00-102.04 mm (4.016-4.017 in.). Then, install a standard piston and ring set.

Cylinder head gasket surface of block should be flat within 0.075 mm (0.003 inch). Surface can be lightly machined to remove imperfections, but distance from centerline of crankshaft main bearings to top surface should not be less than 322.40 mm (12.6929 inches). Standard height of cylinder block, measured from centerline of main bearings to top surface, is 322.90-323.10 mm (12.7126-12.7205 inches).

Cylinder head gaskets of three different thicknesses are available to compensate for cylinder block height. Head gasket with one notch (Fig. 28) should be used on standard height cylinder block.

If 0.25 mm (0.0098 inch) is removed from top surface, cylinder block height will be 322.65-322.85 mm (12.7028-12.7106 inches), the pistons will protrude 0.661-0.910 mm (0.0261-0.0358 inch) and head gasket with two notches should be installed. Cylinder block should be stamped with "XX" mark to indicate that top surface has been machined to this measurement.

If 0.50 mm (0.0197 inch) is removed from top surface, cylinder block height will be 322.40-322.60 mm (12.6929-12.7008 inches), the pistons will protrude 0.911-1.166 mm (0.0359-0.0457 inch) and head gasket with three notches should be installed. Cylinder block should be stamped with "XXX" mark to indicate that top surface has been machined to this measurement.

Top surface of cylinder block should not be machined enough to cause piston protrusion of more than 1.166 mm (0.0457 inch). Machining cylinder head does not affect selection of cylinder head gasket. When assembling, measure the amount of piston stand-out above cylinder block and refer to paragraph 25.

PISTON PINS AND BUSHINGS

All Models

41. The full-floating piston pins (7—Fig. 42) are retained in piston bosses by snap rings and are available in standard size only. Outside diameter of new pin is 39.997-40.003 mm (1.5747-1.5749 in.). If pin is worn to a minimum service limit of 39.990 mm (1.5744 in.), renew pin.

Pin bushing (9) inside diameter in connecting rod should be 40.053-40.067 mm (1.5769-1.5774 in.). If worn to a maximum service limit of 40.092 mm (1.5784 in.), renew connecting rod assembly as bushings are not serviced separately.

Measure piston pin bore in piston, which should be 40.006-40.012 mm (1.5750-1.5753 in.). If worn beyond the maximum service limit of 40.025 mm (1.5758 in.), renew piston.

CONNECTING RODS AND BEARINGS

All Models

42. Connecting rod caps are retained by self-locking cap screws. Connecting rods are removed as outlined in paragraph 38. Standard crankpin diameter is 68.987-69.013 mm (2.716-2.717 in.). Desired bearing clearance is 0.038-0.116 mm (0.0015-0.0045 in.) with a maximum allowable clearance of 0.13 mm (0.005 in.). Regrind or renew crankshaft if out-of-round exceeds 0.05 mm 0.002 in.) or if taper exceeds 0.013 mm 0.0005 in.). Rod bearings are available in undersizes of 0.25, 0.50, 0.75 and 1.0 mm (0.010, 0.020, 0.030 and 0.040 in.) as well as standard. Connecting rod side clearance should be 0.1-0.3 mm (0.004-0.012 in.).

NOTE: If crankpin journals are reground and crankshaft front nose is marked with "N," crankshaft must be rehardened using deep nitroc hardening process.

When reinstalling, stamped cylinder numbers on connecting rods and caps should face opposite camshaft side and arrows on top of pistons must be toward front of engine. Tighten the self-locking rod cap screws on Model 5120 tractors first to 50 N·m (36 ft.-lbs.) and then to 98 N·m (72 ft.-lbs.). On Models 5130 and 5140, tighten connecting rod bolts first to torque of 60 N·m (44 ft.-lbs.), then tighten an additional 60°.

CRANKSHAFT
AND MAIN BEARINGS

All Models

43. On Model 5120, the crankshaft (8—Fig. 43) is supported in five main bearings. Crankshaft end play is controlled by the upper half of flanged No. 4 main bearing insert. On Models 5130 and 5140, the crankshaft (8—Fig. 44) is supported in seven main bearings. Crankshaft end play is controlled by the upper half of flanged No. 4 main bearing insert.

On all models, standard main journal diameter is 82.987-83.013 mm (3.2672-3.2682 in.). Main bearing caps are numbered from No. 1 front to rear and caps must be installed in correct position. Desired main bearing diametral clearance is 0.041-0.119 mm (0.0016-0.0046 in.) with a maximum allowable clearance of 0.140 mm (0.0055 in.). Upper and lower bearing liners are not interchangeable. Make certain the half containing the oil hole is installed in block and that oil holes properly align.

Desired crankshaft end play is 0.137-0.264 mm (0.0054-0.0104 in.). If, after installing new main bearing liners, end play exceeds 0.33 mm (0.013 in.), crankshaft must be renewed. Crankshaft renewal requires removal of engine assembly as outlined in paragraph 24.

Main bearing liners can be renewed after removing oil pan, oil pump intake tube, engine balancer, if so equipped, and main bearing caps. Main bearing lin-

ers are available in undersizes of 0.25, 0.50, 0.75 and 1.0 mm (0.010, 0.020, 0.030 and 0.040 in.) as well as standard.

Main bearing cap screws are self-locking type and can be reused. On Model 5120, tighten main bearing cap screws in three steps. Torque to 60 N•m (44 ft.-lbs.) in step one, 100 N•m (74 ft.-lbs.) in step two and 176 N•m (130 ft.-lbs.) in step three. On Models 5130 and 5140, tighten main bearing cap screws in three steps. Torque to 58 N•m (43 ft.-lbs.) in step one and 80 N•m (59 ft.-lbs.) in step two. In step three, tighten each cap screw an additional 60°.

OIL SPRAY NOZZLES

All Models

44. All engines are equipped with piston cooling spray nozzles located in the main bearing carriers in engine block. The renewable spray nozzles are installed in holes above the upper main bearing halves. Four-cylinder engines have four nozzles and six-cylinder engines are equipped with six nozzles. Spray nozzles need not be removed unless they are damaged. When renewing nozzles, drive nozzles in until flange on nozzle is seated.

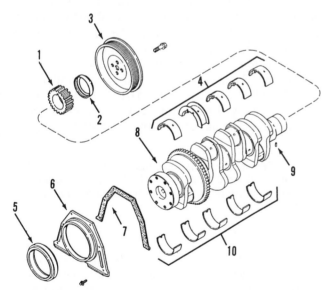

Fig. 43—View of crankshaft and related parts used on Model 5120 engine.

1. Crankshaft gear	6. Seal retainer
2. Front oil seal	7. Gasket
3. Crankshaft pulley	8. Crankshaft assy.
4. Main bearings (upper halves)	9. Pin
5. Rear oil seal	10. Main bearings (lower halves)

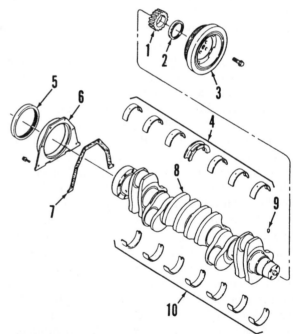

Fig. 44—View of crankshaft and related parts used on Models 5130 and 5140 engines. Center (No. 4) upper bearing half controls crankshaft end play.

1. Crankshaft gear	6. Seal retainer
2. Front oil seal	7. Gasket
3. Crankshaft pulley	8. Crankshaft
4. Main bearings (upper halves)	9. Pin
5. Rear oil seal	10. Main bearings (lower halves)

CRANKSHAFT SEALS

All Models

45. FRONT. To renew the front crankshaft oil seal (2—Fig. 43 or Fig. 44), first remove timing gear cover as outlined in paragraph 31. Then, drive seal from front cover. Thoroughly clean seal and gasket mounting surfaces of front cover and timing gear housing.

> NOTE: The front seal is available only in a kit (Part No. J904353), which includes the two installation tools.

Install two guide studs in the timing gear housing and using a new gasket, slide cover into place. Install cover retaining bolts but do not tighten at this time. Install seal driving tool over crankshaft and into front cover with the small diameter toward the engine. This will center the cover with the crankshaft. Tighten front cover cap screws to a torque of 24 N·m (18 ft.-lbs.). Remove seal driver tool. Place seal on the protective sleeve tool. Apply Loctite 277 to outer edge of seal. Slide protective sleeve and seal on crankshaft until seal contacts the cover. Remove sleeve and install seal driver tool. Drive seal inward until outside diameter of installing tool contacts front cover. This will be the correct installation depth for the seal. Remove seal driver tool. If sealing surface on crankshaft is worn or pitted, a wear ring is available for installation.

For balance of reassembly, refer to paragraph 31.

46. REAR. To renew the rear crankshaft seal (5— Fig. 43 or Fig. 44), first split tractor between engine and speed transmission as outlined in paragraph 24. Remove screws and lift out baffle plate (1—Fig. 45). Unbolt and remove the torque limiter (15, 16 and 17) from flywheel (13), then unbolt and remove flywheel. Drain engine oil and remove oil pan. Attach a hoist to flywheel housing. Unbolt and remove starter motor. Remove cap screws (11) and lift off flywheel housing (2). Remove "O" ring (10) from around the rear seal retainer (6—Figs. 43 or 44). Unbolt and remove seal retainer and oil seal, then remove seal from retainer. Clean gasket mounting surfaces of retainer and crankcase. Use Loctite safety solvent and clean mounting surface of retainer seal.

> NOTE: The rear crankshaft seal is available only in a kit (Part No. J909411 for Model 5120 or Part No. J909410 for Models 5130 and 5140), which includes two installation tools.

Install two guide studs in rear of crankcase. Using a new gasket (7), slide seal retainer in place over guide studs. Remove guide studs, install, but do not tighten, retainer bolts. Install seal driver tool into seal retainer with small diameter toward the engine.

This will center the retainer with the crankshaft. Tighten seal retainer cap screws to a torque of 9 N·m (7 ft.-lbs.). Remove the seal driver tool. Slide protective sleeve tool with seal over crankshaft until outer diameter of seal contacts seal retainer. Remove sleeve and install seal driver tool. Drive seal inward until outside diameter of installing tool contacts seal retainer. This will be the correct installation depth for the seal. Remove seal driver tool. If sealing surface on crankshaft flange is worn or pitted, a wear ring is available for installation.

Install new "O" ring (10—Fig. 45) over seal retainer. Apply a bead of Loctite 515 to the sealing surface of engine block. Install two guide studs in

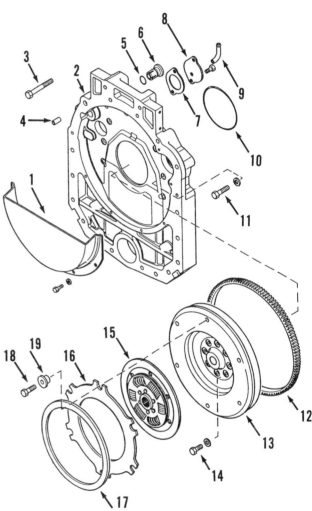

Fig. 45—Flywheel housing and related parts used on all models.

1. Oil baffle	
2. Flywheel housing	11. Cap screw (8)
3. Cap screw	12. Flywheel ring gear
4. Dowel	13. Flywheel
5. "O" ring	14. Cap screw (8)
6. Plug	15. Clutch plate
7. Gasket	16. Pressure plate
8. Cover	17. Belleville spring
9. Tube	18. Cap screw (6)
10. "O" ring	19. Retainer (6)

engine block, then carefully install flywheel housing. Install flywheel housing bolts (11) and remove guide studs. Tighten flywheel housing bolts to a torque of 60 N•m (45 ft.-lbs.). Install flywheel and tighten bolts (14) to a torque of 137 N•m (101 ft.-lbs.). Install torque limiter clutch and tighten bolts (18) to a torque of 134-151 N•m (100-110 ft.-lbs.). Reinstall oil baffle plate (1). The balance of reassembly is the reverse of disassembly procedure. Refer to paragraph 24 and reconnect tractor.

FLYWHEEL

All Models

47. To remove the flywheel (13—Fig. 45), first split tractor between engine and speed transmission housing as outlined in paragraph 24. Remove oil baffle plate (1) and torque limiter clutch, then unbolt and remove flywheel.

The flywheel ring gear (12) is a shrink fit on front of flywheel. The manufacturer cautions against the use of a torch or open flame for heating ring gear. Heat the new ring gear (12) to a temperature of 204-232° C (400-450° F) in an oven. Install gear on flywheel.

Install flywheel and tighten the retaining cap screws (14) to a torque of 137 N•m (101 ft.-lbs.). Reassemble tractor in reverse order of disassembly.

ENGINE BALANCER

Model 5120

48. Four-cylinder engines are equipped with an engine balancer. Two types of balancers have been used and are identified as a two-gear balancer or as a three-gear balancer. The two-gear balancer is driven by a gear as shown on crankshaft (8—Fig. 43). The three-gear balancer is also driven by the gear on crankshaft (8) and an idler on the balancer. To remove either balancer, first remove the oil pan and oil pick-up tube. Unbolt and remove balancer from engine crankcase.

49. TWO-GEAR BALANCER. To disassemble the balancer, unbolt and remove oil tube (1—Fig. 46). Then, unbolt and remove weights (5) from shafts (8). Withdraw gear and shaft assemblies from housing (4). Spacers (6) can be removed as shafts are withdrawn.

Inspect shafts (8) and gears (7) for excessive wear or other damage. Shafts and gears are available separately and gears can be pressed on shafts to position shown in Fig. 47.

> NOTE: The gears must be pressed onto the shafts in the correct position so the balancer can be timed. If precision measuring tools are not available to facilitate correct installation of gears, renew gear and shaft as assemblies.

Inspect bushings (12—Fig. 46) for wear or scoring and renew as necessary. Bushings and expansion plugs can be pressed from housing (4).

Reassemble balancer by reversing the disassembly procedure. Tighten weight bolts (9) to a torque of 24-32 N•m (18-24 ft.-lbs.). With weights at bottom, make certain the timing marks on gears are aligned as shown in Fig. 48. With timing marks aligned, screw timing lock bolt (11—Fig. 46) up through bottom of housing (4) and into slot in balancer weight. This will hold balancer in time during installation.

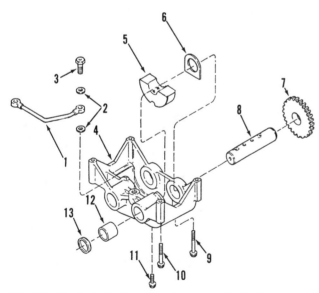

Fig. 46—Exploded view of two-gear engine balancer used on some Model 5120 tractors.

1. Oil tube
2. Copper washers
3. Cap screw
4. Housing
5. Balancer weight (2)
6. Spacer (2)
7. Gear (2)
8. Shaft (2)
9. Weight bolt (4)
10. Mounting bolt (4)
11. Timing lock bolt
12. Bushing (4)
13. Expansion plug (2)

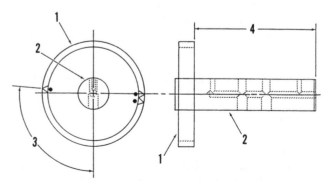

Fig. 47—Correct position for installing gear on shaft.

1. Gear
2. Shaft
3. 94.3-94.9°
4. 4.92-4.96 in. (125-126 mm)

Use flywheel rotating tool to turn engine crankshaft in normal direction of rotation while pressing inward on timing pin. When timing pin engages hole in camshaft gear and No. 1 piston is at TDC, engine will be in position for balancer installation. Use new copper washers (2) and install oil tube (1). Install balancer assembly and tighten mounting bolts (10) to a torque of 21-27 N·m (15-20 ft.-lbs.). **Remove timing lock bolt (11), disengage timing pin and rotate engine to see that balancer rotates freely.** Remove flywheel rotating tool, install oil pick-up tube and oil pan. Fill crankcase with correct amount of engine oil.

50. THREE-GEAR BALANCER. To disassemble the three-gear balancer, remove the Allen head bolts (11—Fig. 49) and remove gear support (12), idler drive gear (13) and needle bearing (14). Unbolt and remove end cap (22). Remove weight bolts (3), then remove upper and lower weights (2 and 7). Press shafts out of thrust collars and remove shafts from rear end of housing (19). Remove thrust collars (21) and thrust plate (20). Separate idler gear (13) from support (12) and needle bearing (14) from idler gear.

Inspect shafts (1 and 6), gears (16) and bearing sleeves (9). If necessary, press gears from shafts and remove Woodruff keys (4). To remove bearing sleeves (9), drive roll pins (5) into shafts, then press sleeves from shafts. Inspect needle bearings (8 and 10) in housing (19). Press damaged needle bearings from housing.

Refer to Fig. 50 for correct placement of new bearings. Rear (gear end) bearings should be installed 0.20-0.33 mm (0.008-0.013 in.) below flush with rear face of housing. Center bearing should be installed 176.07-176.20 mm (6.93-6.94 in.) from rear face of housing (with oil holes aligned). Front bearing should be installed 3.37-3.50 mm (0.133-0.138 in.) below thrust plate surface of housing.

Reassemble by reversing the disassembly procedure. Refer to Figs. 49 and 50 as a guide during reassembly. Install single weights (2) on upper shaft (1) and double weights (7) on lower shaft (6). Tighten weight bolts to a torque of 21-27 N·m (15-20 ft.-lbs.). Tighten the four end cap bolts (23) to 21-27 N·m (15-20 ft.-lbs.). Install idler drive gear (13), needle bearing (14) and gear support (12—Fig. 49), then tighten Allen head bolts (11) to a torque of 52-62 N·m (38-46 ft.-lbs.).

With balancer assembled, align timing marks on gears (16) as shown in Fig. 51. With marks aligned, install timing lock bolt (17—Figs. 49 or 50) through bottom of housing (19) and into hole in lower shaft (6). This will hold balancer in time during installation. Use flywheel rotating tool to turn engine crankshaft in normal direction of rotation while pressing inward on timing pin. When timing pin engages hole in camshaft gear and No. 1 piston is at TDC, engine

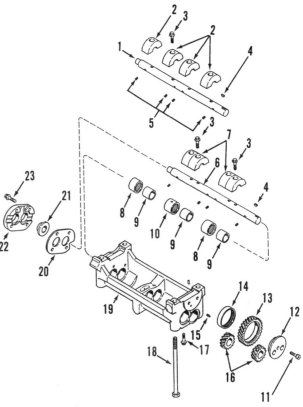

Fig. 49—Exploded view of three-gear engine balancer used on some Model 5120 tractors.

1. Upper shaft
2. Upper balancer weights
3. Weight bolts (8)
4. Woodruff keys
5. Roll pins
6. Lower shaft
7. Lower balancer weights
8. End needle bearings (4)
9. Bearing sleeves
10. Center needle bearings (2)
11. Bolt (2)
12. Gear support
13. Idler drive gear
14. Needle bearing
15. Roll pin
16. Shaft gears
17. Timing lock bolt
18. Mounting bolt (4)
19. Housing
20. Thrust plate
21. Thrust collar (2)
22. End plate
23. Bolt (4)

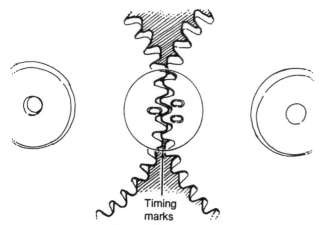

Timing marks

Fig. 48—Timing marks aligned on two-gear engine balancer during assembly.

will be in position for installation of balancer. Install balancer assembly and tighten mounting bolts (18—Fig. 49) to a torque of 167-183 N·m (123-135 ft.-lbs.). **Remove timing lock bolt (17), disengage timing pin and rotate engine crankshaft to see that balancer rotates smoothly.** Remove flywheel rotating tool, install oil pick-up tube and oil pan. Fill crankcase with correct amount of engine oil.

OIL PAN AND
OIL INLET TUBE

All Models

51. To remove the oil pan, remove plug and drain the oil. On models equipped with front-wheel drive, remove front drive shaft shield and the drive shaft. Unbolt and remove the oil pan, then unbolt and remove oil inlet tube.

Clean all surfaces of foreign material. Using new gasket, install oil inlet tube. Tighten retaining cap screws to a torque of 24 N·m (18 ft.-lbs.). Install oil pan with new gasket and tighten pan bolts to a torque of 24 N·m (18 ft.-lbs.).

If so equipped, install front drive shaft and shield as outlined in paragraph 9. Fill crankcase with new oil to proper level. Refer to CONDENSED SERVICE DATA for capacities.

OIL PUMP

All Models

52. To remove the oil pump, first remove timing gear cover as outlined in paragraph 31. With front cover removed, unbolt and remove oil pump assembly as shown in Fig. 52. Place oil pump on a bench with gear side down. Remove rear cover plate. Then, using a feeler gauge, measure clearance between lobe on inner rotor and lobe on outer rotor as shown in Fig. 53. If clearance exceeds 0.18 mm (0.007 in.), renew oil pump assembly. Oil pump is serviced only as a complete assembly.

When installing oil pump assembly, tighten the four mounting bolts evenly to a torque of 24 N·m (18 ft.-lbs.) following the sequence shown in Fig. 54. Do not overtighten. A small clearance will exist between pump flange and cylinder block.

With pump installed, check backlash between the pump idler gear and the crankshaft gear. If backlash exceeds 0.45 mm (0.018 in.), renew pump assembly.

Refer to paragraphs 31 and 45, and reassemble tractor.

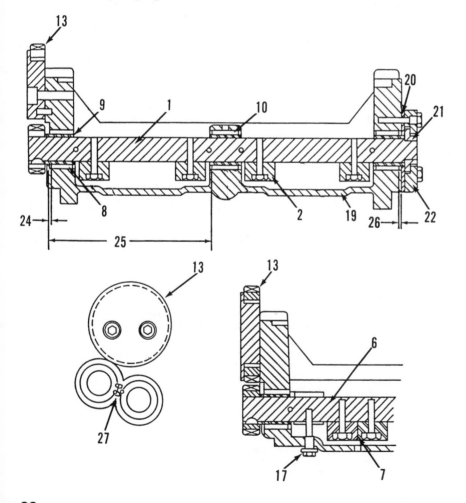

Fig. 50—Cut-away views for aid in reassembly of three-gear balancer, showing bearing placement in housing.

1. Upper shaft
2. Upper balancer weights
6. Lower shaft
7. Lower balancer weights
8. End needle bearings
9. Bearing sleeves
10. Center needle bearing
13. Idler drive gear
17. Timing lock bolt
19. Housing
20. Thrust plate
21. Thrust collar
22. End plate
24. 0.20-0.33 mm (0.008-0.013 in.)
25. 176.07-176.20 mm (6.93-6.94 in.)
26. 3.37-3.50 mm (0.133-0.138 in.)
27. Timing marks

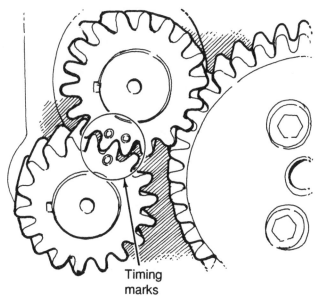

Fig. 51—Timing marks aligned on three-gear engine balancer during assembly.

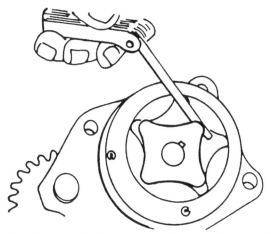

Fig. 53—Using a feeler gauge, measure clearance between lobe on inner rotor and lobe on outer rotor. Refer to text.

Fig. 52—With timing gear front cover removed, unbolt and remove oil pump assembly.

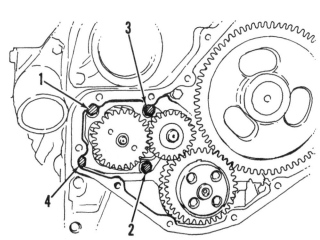

Fig. 54—Use tightening sequence shown when installing oil pump.

OIL PRESSURE REGULATOR

All Models

53. The oil pressure regulator is located in the oil filter head (5 or 11—Fig. 55). Two types of filter heads have been used: one with bottom-mounted regulator and the other with top-mounted regulator. Remove plug (9), gasket (8), spring (7 or 12) and plunger (6 or

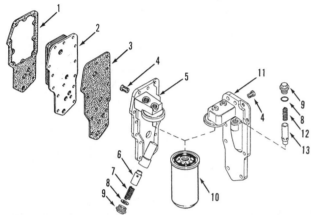

Fig. 55—Exploded view of engine oil cooler, oil filter head and relative components. Note different locations of regulator valves.

1. Gasket	8. Gasket
2. Oil cooler	9. Plug
3. Gasket	10. Filter cartridge
4. Cold oil relief valve	11. Filter head (upper
5. Filter head (lower	regulator valve)
regulator valve)	12. Regulator valve
6. Regulator plunger	spring
(two-hole)	13. Regulator plunger
7. Regulator valve spring	(four-hole)

13). Free length of spring (7) should be 55.83 mm (2.198 in.) and should test 95-113 N (21.4-25.4 lbs.) when compressed to a length of 39.98 mm (1.57 in.). Spring (12) should have a free length of 64.0 mm (2.52 in.) and should test 104.5 N (23.5 lbs.) when compressed to a length of 41.25 mm (1.62 in.).

When reinstalling, use new gasket (8) and tighten plug securely. Oil pressure on new engine, with engine warm and operating at low idle speed, should be 193-296 kPa (28-43 psi) and at 2400 rpm should be 296-372 kPa (43-54 psi). Oil pressure on used engine, with engine warm and operating at low idle speed, should be a minimum of 69 kPa (10 psi) and at 2400 rpm should be at least 207 kPa (30 psi).

ENGINE OIL COOLER

All Models

54. The engine oil cooler (2—Fig. 55) is located in right side of engine block behind the oil filter head. To remove the oil cooler, drain cooling system and remove the oil filter cartridge. If so equipped, disconnect the turbocharger oil supply line from filter head. Unbolt and remove filter head (5 or 11) and gasket (3). Remove oil cooler (2) and gasket (1).

Clean all surfaces of foreign material. Cleaning procedures for the oil cooler consists of flushing the oil circuit with clean solvent and removing scale or rust from exterior of plates.

Using new gaskets, reinstall by reversing the removal procedure. Make certain the cold oil relief valve (4) is in position in backside of filter head. Tighten filter head retaining bolts to a torque of 24 N·m (18 ft.-lbs.).

DIESEL FUEL SYSTEM

Models 5120 and 5140 are equipped with Robert Bosch Model VE injection pumps and Model 5130 is equipped with a CAV Model DPA injection pump. Robert Bosch multi-hole injection nozzles are used on all models.

FUEL, FUEL FILTERS AND LIFT PUMP

All Models

55. FUEL. The manufacturer recommends the use of a good grade No. 2 diesel fuel having a Cetane rating of 40 or better be used for normal operation. For cold weather operation, Cetane rating should be 45-55 and fuel pour point should be at least –12° C (10° F) below lowest expected ambient temperature.

56. FUEL LIFT PUMP. A diaphragm-type fuel lift pump (7—Fig. 56) is located on left side of engine block. This unit pumps the fuel through the two filters and on to the fuel injection pump. The lift pump is operated by a lobe on engine camshaft. A hand lever used for bleeding air from fuel system if located on the lift pump. The pump is serviced only as a complete unit.

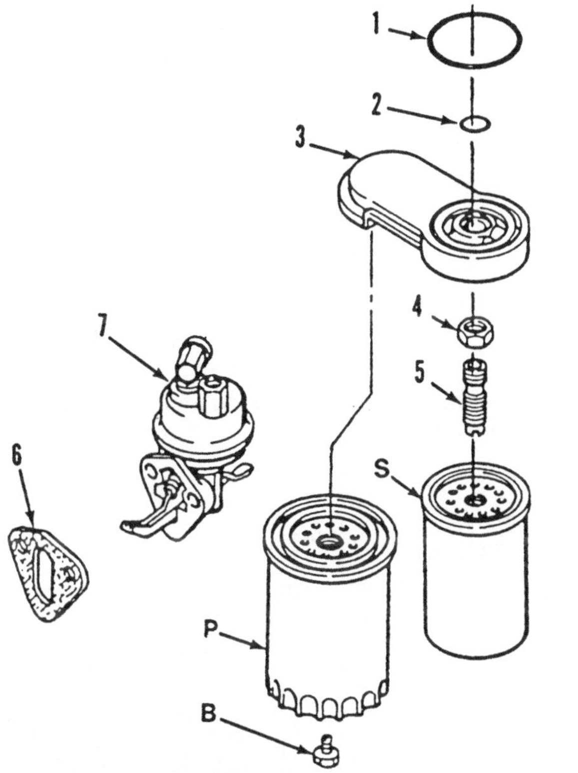

57. FUEL FILTERS. Filters are the screw-on type (Fig. 56). Filters are available in a kit, which consists of both units and both should be renewed each 500 hours of operation or when noticeable power loss occurs. Primary filter (P—Fig. 56 or Fig. 57) is equipped with a water bleed screw (B). Each morning before operating tractor, loosen water bleed screw and drain off any water from primary filter.

To renew filters, first clean filter head, elements and surrounding area. Remove water bleed screw (B) from primary filter (P). To remove the fuel filters, use a clamp-type filter wrench.

Lightly oil gaskets on new units and turn filters on until gaskets contact filter head. Then, hand-tighten filters ½ to ¾ of a turn. DO NOT overtighten. Bleed system as outlined in paragraph 58.

58. BLEED SYSTEM. To bleed the fuel system, make certain there is fuel in the tank. Loosen air vent screw (V—Fig. 57) and operate hand lever on fuel lift pump until air-free fuel flows from vent screw opening. Close air vent screw (V) and loosen vent screw on rear of Robert Bosch pump or on side of CAV pump. Turn ignition switch to ON position to energize the solenoid shut-off in injection pump. Again, operate hand lever on fuel lift pump until air-free fuel flows from vent screw opening. Close vent screw and operate hand lever several additional strokes.

Fig. 56—Screw-on-type fuel filters are used on all tractors. Filter head (3) is secured to flange on cylinder head by adapter tube (5) and nut (4).

B. Water bleed screw	3. Filter head
P. Primary filter	4. Nut
S. Secondary filter	5. Adapter tube
1. "O" ring	6. Gasket
2. "O" ring	7. Fuel lift pump

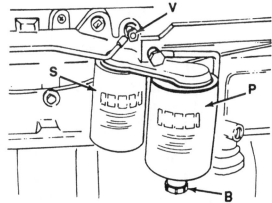

Fig. 57—Screw filters on until seals contact filter head, then hand tighten filters ½ to ¾ of a turn.

B. Water bleed screw	S. Secondary filter
P. Primary filter	V. Air vent screw

INJECTION PUMP

All Models

59. LUBRICATION. The fuel injection pump is lubricated with diesel fuel under 34.5-48.3 kPa (5-7 psi) pressure. Excess fuel flows through the return line where it is joined by injector leak-off fuel and returns to the fuel tank.

60. TIMING (MODEL 5130). The injection pump drive shaft is keyed to the pump drive gear. As long as the pump drive gear is in proper relation to the engine timing gear train, as outlined in paragraph 34, and the timing mark on pump mounting flange is aligned with the mark on timing gear housing, as shown in Fig. 58, pump timing is correct.

If pump timing marks are not in alignment, loosen pump bracket bolt and the three pump mounting nuts. Rotate pump assembly to align the marks, then tighten mounting nuts and bracket bolt to a torque of 24 N·m (18 ft.-lbs.).

61. TIMING (MODELS 5120 AND 5140). Injection pump timing can be considered correct if timing mark on pump mounting flange is aligned with timing mark on timing gear housing (Fig. 58A). If the accuracy of the timing marks is questioned, pump timing can be checked and adjusted as follows: Install flywheel rotating tool (Fig. 59), push inward on timing pin and rotate flywheel clockwise (viewed from front) until timing pin enters hole in camshaft gear. Remove injection lines and vent plug from rear of injection pump. Install a dial indicator (CAS-10066-1A) and adapter (CAS-1745) as shown in Fig. 60. Pull timing pin out of the camshaft gear and rotate flywheel counterclockwise (viewed from front) until there is no movement on the dial indicator. Set indicator dial at zero. Push inward on timing pin and rotate flywheel clockwise until timing pin enters hole in camshaft gear. At this time, the dial indicator should read 1.5 mm (0.059 in.) if timing is correct. If timing is not correct, loosen pump rear brace bolt and the three nuts securing pump to timing gear housing.

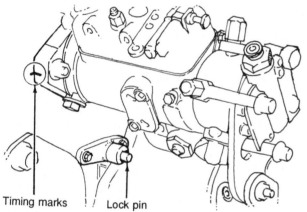

Timing marks   Lock pin

Fig. 58—CAV injection pump used on Model 5130 is in time if timing marks on pump flange and timing gear housing are aligned.

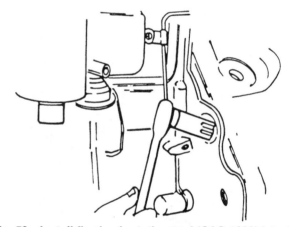

Fig. 59—Install flywheel rotating tool (CAS-1690) into flywheel housing.

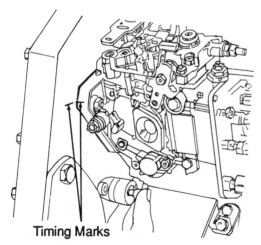

Timing Marks

Fig. 58A—View of timing marks on Robert Bosch injection pump and timing gear housing. Timing can be considered correct when marks are aligned.

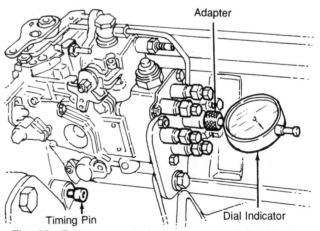

Timing Pin   Dial Indicator

Fig. 60—Remove vent plug from rear of Robert Bosch injection pump and install dial indicator (CAS-10066-1A) and adapter (CAS-1745) as shown.

Rotate pump assembly until dial indicator reads 1.5 mm (0.059 in.). Then, tighten pump mounting nuts and brace bolt to a torque of 24 N·m (18 ft.-lbs.).

NOTE: If correct reading cannot be obtained, timing gears are not correctly timed.

Remove dial indicator and adapter from pump. Install vent plug and tighten to a torque of 8-10 N·m (6-7 ft.-lbs.). Install fuel injection lines. Disengage timing pin from camshaft gear. Remove flywheel rotating tool and install plug in flywheel housing. Bleed fuel system as outlined in paragraph 58.

62. REMOVE AND REINSTALL. To remove the CAV or Robert Bosch injection pump, first thoroughly clean pump, lines, connections and surrounding area. Remove fuel injection lines, inlet line and excess fuel line from pump. Immediately cap or plug all openings. Remove tachometer drive or access plug from left front side of front cover and remove nut and washer from front end of pump shaft. Remove plug from flywheel housing and install flywheel rotating tool (CAS-1690). See Fig. 59. Push inward on timing pin while rotating flywheel in normal direction of rotation until timing pin engages hole in camshaft gear. On CAV pump, loosen lock bolt and move slot washer upward as shown in Fig. 61. Then, tighten lock bolt to a torque of 30 N·m (22 ft.-lbs.). Washer will be loose in lock position with bolt tightened. On Robert Bosch pumps, loosen lock bolt and remove slotted tab (Fig. 62), then tighten shaft lock bolt. On all models, loosen pump rear brace bolt and the three nuts securing pump to timing gear housing several turns. Install gear puller (CAS-1691A) as shown in Fig. 63 and pull gear from taper on pump shaft. Remove brace bolt and the three pump mounting nuts. Carefully remove injection pump assembly.

CAUTION: If the Woodruff key is not in the pump shaft when pump is removed, engine front cover must be removed to retrieve the key.

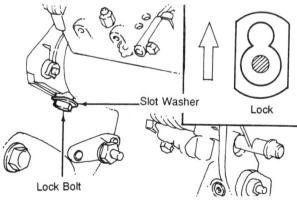

Fig. 61—On CAV pump, loosen shaft lock bolt, move slot washer up to lock position, then retighten lock bolt.

Using a new pump mounting gasket, reinstall pump by reversing the removal procedure. Before installing the injection lines, set pump timing as outlined in paragraph 60 or 61. Tighten pump drive gear nut to a torque of 65 N·m (48 ft.-lbs.). Tighten pump mounting nuts and brace bolt to a torque of 24 N·m (18 ft.-lbs.). When moving slot washer down in unlock position (CAV pump), or installing slotted tab (Robert Bosch pumps), tighten lock bolt to a torque of 30 N·m (22 ft.-lbs.).

If a new pump is being installed that does not have a timing mark on pump flange, scribe a mark on the flange in line with timing mark on timing gear housing (Figs. 58 and 58A) for future timing reference.

If so equipped, use new gaskets on each side of fuel lines and tighten banjo-fitting bolts to 24 N·m (18 ft.-lbs.). Bleed fuel system as outlined in paragraph 58.

63. SPEED ADJUSTMENT. The diesel governor is an integral part of injection pump and, except for minor speed adjustments, governor service should be done at an authorized diesel service shop.

Recommended governed speeds for all models are as follows:

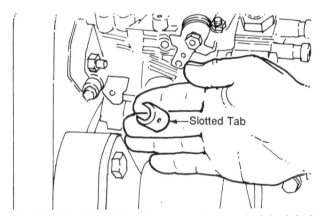

Fig. 62—On Robert Bosch pumps, loosen shaft lock bolt, remove slotted tab and retighten lock bolt.

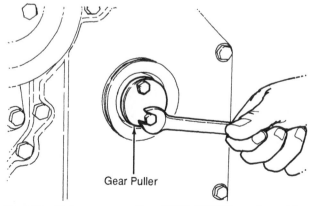

Fig. 63—Using gear puller (CAS-1691A) and working through access hole in front cover, pull gear from pump shaft.

Low idle . . . . 900 rpm
High idle (no-load) . . . . 2400 rpm
Rated (loaded) speed . . . . 2200 rpm

To adjust governed speeds, first make certain that throttle cable is correctly adjusted to fully open and

close governor control lever. Turn stop screws (6 and 11—Fig. 64 or Fig. 65) to adjust high and low idle speeds.

64. FUEL SHUT-OFF SOLENOID. To remove the fuel shut-off solenoid (7—Fig. 64 or Fig. 65),

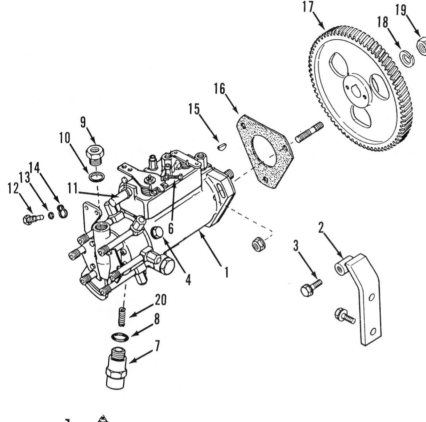

Fig. 64—CAV diesel injection pump and relative components used on Model 5130.

1. Injection pump
2. Pump rear brace
3. Brace bolt
4. Vent plug
6. Low-idle stop screw
7. Fuel shut-off solenoid
8. Seal
9. Fuel inlet fitting
10. Gasket
11. High-idle stop screw (under cap)
12. Pump shaft lock bolt
13. "O" ring
14. Slot washer
15. Woodruff key
16. Gasket
17. Pump drive gear
18. Lockwasher
19. Nut
20. Spring

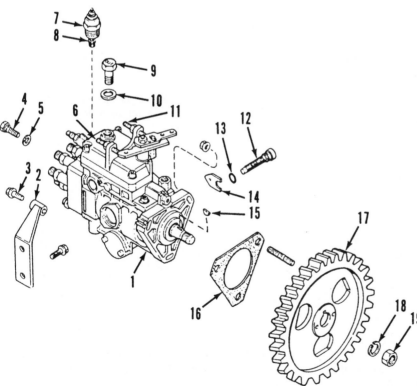

Fig. 65—Robert Bosch diesel injection pump and relative components used on Model 5140. Robert Bosch pump used on Model 5120 is similar.

1. Injection pump
2. Pump rear brace
3. Brace bolt
4. Vent plug
5. Gasket
6. Low-idle stop screw
7. Fuel shut-off solenoid
8. Seal
9. Banjo bolt
10. Gasket
11. High-idle stop screw
12. Pump shaft lock bolt
13. "O" ring
14. Slotted tab
15. Woodruff key
16. Gasket
17. Pump drive gear
18. Lockwasher
19. Nut

disconnect electrical wire at connector. Unscrew and remove solenoid and seal (8). Refer to Fig. 66 and disassemble solenoid. Inspect plunger and solenoid for scoring or other damage. Make certain the orifice in plunger is free of foreign material.

Reassemble solenoid and, using new "O" ring, reinstall unit. Tighten solenoid to a torque of 15 N·m (11 ft.-lbs.).

INJECTOR NOZZLES

All Models

Both 7 mm and 9 mm Robert Bosch injectors have been used. Refer to the following:

Engines manufactured in
Darlington, England:
    Prior to engine serial No. 21092870 . . . . 9 mm
    Engine serial No. 21092870 and after . . . . . 7 mm
Engines manufactured in
Neuss, Germany:
    Prior to engine serial No. 52107489 . . . . 9 mm
    Engine serial No. 52107489 and after . . . . . 7 mm
Engines manufactured in
North America:
    Prior to engine serial No. 44511034 . . . . 9 mm
    Engine serial No. 44511034 and after . . . . . 7 mm

See Fig. 67 for a cross-sectional view of injector nozzle.

65. LOCATING A FAULTY NOZZLE. If rough or uneven engine operation or misfiring indicate a faulty injector, defective unit can usually be located as follows:

With engine operating at speed where malfunction is most noticeable (usually slow idle speed), loosen compression nut on high-pressure line at each injector in turn and listen for change in engine performance. As in checking spark plugs, faulty unit is the one that, when its line is loosened, least affects engine operation.

If a faulty nozzle is found and considerable time has elapsed since injectors have been serviced, it is recommended that all nozzles be removed and checked or that new or reconditioned units be installed. Refer to appropriate following paragraphs for removal and test procedures.

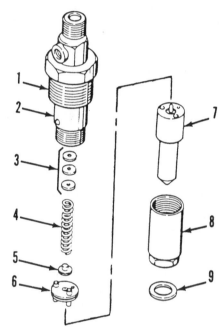

Fig. 68—Exploded view of Robert Bosch diesel fuel injector.

1. Injector nut
2. Injector body
3. Shims
4. Spring
5. Spring seat
6. Valve stop
7. Valve assy.
8. Cap nut
9. Injector seal

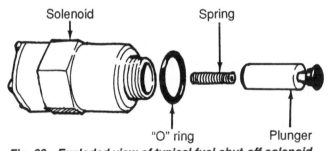

Fig. 66—Exploded view of typical fuel shut-off solenoid.

Fig. 67—Cross-sectional view of diesel fuel injector assembly.

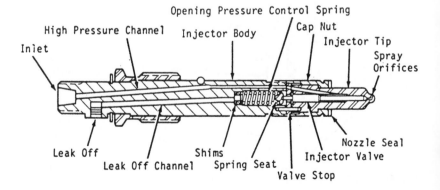

66. REMOVE AND REINSTALL. Clean the area around the injectors. Unbolt injection lines bracket. Loosen injection line tube nut at pump, then disconnect line from injector. Unbolt leak-off line from injector. Cap or plug all openings. Using special tool (CAS-1066A) on injector nut (1—Fig. 68) and a wrench on the flat surfaces of injector body (2) to prevent injector from turning, unscrew nut and remove injector.

Make certain that injector seal (9) is removed with injector. If not, use suitable tool or wire to remove the seal from the bore. Remove the remaining injectors using the same procedure. Clean bores for injectors using special tool (CAS-1694) for 9 mm injectors or (CAS-2155) for 7 mm injectors.

Using a small amount of petroleum jelly to hold a new seal (9) in place, reinstall injector. The ball in the injector must align with the slot in the bore. Tighten injector nut (1) to a torque of 60 N·m (44 ft.-lbs.). Install leak-off line and tighten bolt to a torque of 8 N·m (6 ft.-lbs.). Install and tighten injection lines.

Start engine. If engine fails to start, loosen injection line nuts at injectors. Turn ignition switch on and crank engine until diesel fuel leaks at loosened connections. Tighten line nuts and start engine.

67. TESTING. A complete job of testing and adjusting an injector nozzle requires use of special test equipment, such as test stand (CAS-10091) shown in Fig. 69. Only clean diesel fuel or approved testing oil should be used in tester tank. The nozzle should be tested for opening pressure, seat leakage and spray pattern. When tested, nozzle should open with a sharp popping or buzzing sound and cut off quickly at end of injection with a minimum of seat leakage and a controlled amount of back leakage. Check as outlined in the following paragraphs.

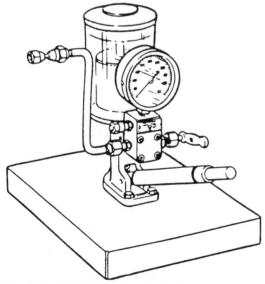

Fig. 69—CAS-10091 diesel fuel injection nozzle tester.

WARNING: Fuel leaves nozzle tip with sufficient force to penetrate the skin. Keep hands away from nozzle tip when testing.

68. OPENING PRESSURE. Before conducting test, operate tester until fuel flows, then attach injector to tester using correct adapter. Close valve to tester gauge and pump tester lever a few quick strokes to be sure that nozzle is not plugged and that all four spray holes are open.

Open valve to tester gauge and operate tester until fuel flows, then observe gauge reading. Opening pressure should be 24,500-25,305 kPa (3553-3670 psi). If opening pressure is not a specified, disassemble injector as outlined in paragraph 71 and vary shims (3—Fig. 68) as required. Shims are available in thirty thicknesses ranging from 1.0 mm (0.039 in.) to 1.98 mm (0.078 in.). The opening pressure is changed about 380 kPa (55 psi) for each 0.025 mm (0.001 in.) of shim thickness that is added or removed.

NOTE: When testing all the injectors in an engine, there must not be more than 1035 kPa (150 psi) difference between any of the injectors.

69. SPRAY PATTERN. Close the valve to tester gauge and operate tester handle at approximately 60 strokes per minute. Four finely atomized conical sprays should be emitted from injector tip. If spray is solid, streaked or irregular, disassemble and overhaul injector as outlined in paragraph 71.

70. SEAT LEAKAGE. Point injector tip downward and operate tester rapidly to be sure nozzle valve is seated. Dry injector tip, then operate tester slowly to raise gauge pressure to within 690 kPa (100 psi) under the opening pressure and maintain this pressure for five seconds. Observe the injector tip and if nozzle is in good condition, there should be no fuel accumulation in the five second period, although a slight dampness is permissible. If fuel drips or a visible drop forms, renew injector or overhaul as outlined in paragraph 71.

71. OVERHAUL. Soak injector in a carbon removing solvent and clean outside of injector thoroughly. Clamp flats of injector body (2—Fig. 68) in a soft-jawed vise so injector tip is pointing upward. Use a box wrench to remove cap nut (8) and be sure nozzle valve (7) does not turn with cap nut. If nozzle valve does turn with cap nut, dowels and/or valve stop (6) will be damaged. Return injector assembly to solvent if nozzle valve is stuck to cap nut.

With cap nut removed, nozzle valve (7), valve stop (6), spring seat (5), spring (4) and shims (3) can be removed from body (2). Note number and thickness of shims (3) removed for aid in reassembly. Shims are available in thirty thicknesses ranging from 1.0 mm (0.039 in.) to 1.98 mm (0.078 in.).

Nozzle valve assembly (7) is a matched set and parts should never be intermixed. Keep parts for each injector separate and immersed in clean diesel fuel or solvent in a compartmented pan as injector is disassembled.

A brass wire brush can be used to clean exterior surfaces of nozzle valve (7). Clean nozzle spray orifices using cleaning wires 0.013-0.025 mm (0.0005-0.001 in.) smaller than orifice holes. Orifice holes are 0.28-0.29 mm (0.0110-0.0115 in.). Each nozzle contains four spray holes. Use compressed air to clean ducts in nozzle and a reverse flow of air through tip end of nozzle is recommended. Clean nozzle valve seat area by using a brush and, if necessary to remove varnish, use solvent and a felt cleaning pad. Make sure that nozzle valve slides freely in its body. Renew any parts that are scratched, scored, eroded or show signs of any other damage.

Reassemble injector as follows: Be sure parts are wet with diesel fuel and install adjusting shims (3), spring (4) and spring seat (5) with small diameter side toward spring. Align dowel pins in valve stop (6) with dowel holes in body (2) and install valve stop. Install valve assembly (7) on valve stop, apply Lubriplate to shoulder only of valve assembly, then install cap nut (8). Clamp flats of injector body in a soft-jawed vise and tighten cap nut to a torque of 24 N·m (18 ft.-lbs.).

Recheck injector opening pressure and, if necessary, readjust opening pressure as outlined in paragraph 68.

TURBOCHARGER

A Holset turbocharger is used on Models 5120 and 5140. The unit mounts on exhaust manifold and uses exhaust pressure to slightly compress the intake air.

Models 5120-5140

72. OPERATION. Exhaust gas energy turns the turbine wheel at a speed that may vary from a few thousand to 125,000 rpm, depending on engine speed and load. The compressor wheel supplies air to the intake manifold at or above normal atmospheric pressure. The additional air entering the combustion chamber permits an increase in the amount of fuel that can be burned, and increases the power output over an engine of comparable size not so equipped.

The use of engine exhaust to power the compressor increases engine flexibility, enabling it to perform with the economy of a small engine on light loads, yet permitting substantial horsepower increase at full load. Horsepower loss because of altitude or atmospheric pressure changes is also largely eliminated.

All diesel engines operate with an excess of air under light loads. In a naturally aspirated engine, most of the air is used at full load and increasing the amount of fuel results in a higher exhaust smoke level with little increase in power output. Turbocharging provides a variation of air delivery and a turbocharged engine operates with an excess of air up to or beyond the designed capacity of the engine. When more fuel is provided, the turbocharger speed and air delivery increase, resulting in additional horsepower and heat with little change in smoke level. Exhaust smoke cannot, therefore, be used as an accurate guide to maximum fuel setting for a turbocharged engine.

The turbocharger turbine shaft bearings are floating sleeve-type, and unit is lubricated and cooled by a flow of engine oil under pressure.

73. REMOVE AND REINSTALL. To remove the turbocharger, first raise hood and proceed as follows: Loosen clamps and remove turbocharger intake elbow. Remove inner seal clamp and exhaust elbow retaining bolts. Lift off elbow. Loosen clamps and remove air outlet hose and tube. Disconnect oil supply line and oil drain line from turbocharger. Unbolt and remove turbocharger from exhaust manifold.

When reinstalling, use a new gasket and tighten turbocharger mounting nuts to a torque of 32 N·m (24 ft.-lbs.). Install new gasket on oil drain line and tighten bolts to a torque of 24 N·m (18 ft.-lbs.). Install oil supply tube and tighten securely. Complete installation by reversing the removal procedure.

Turbocharger MUST be primed after installation as follows: Disconnect plug for fuel shut-off solenoid wire. Turn key switch and crank engine for approxi-

mately 30 seconds. Reconnect solenoid wire and set speed control lever to one-third speed position. Start engine and operate at 1000 rpm for about two minutes to supply normal lubrication to the turbocharger. Then, increase engine speed to 1800 rpm until normal engine oil pressure indicator assures proper lubrication to the turbocharger.

74. OVERHAUL. With turbocharger removed as outlined in paragraph 73, place match marks across compressor housing (1—Fig. 70) and back plate (4), across back plate and center housing (16) and across center housing and turbine housing (22) for aid in correct reassembly.

> **CAUTION: Do not rest weight of any parts on compressor wheel or turbine wheel blades. Weight of only the turbocharger unit is enough to damage the blades.**

Remove cap screws (12) with washers and clamping plates (13), then lift off compressor housing (1). Straighten tabs of lockplates (24) and remove cap screws with lockplates and clamping plates (23). Lift center housing assembly from turbine housing (22). Place turbine wheel end of center housing assembly in a holding fixture to prevent turbine shaft from turning and remove nut (2).

> **NOTE: Use a "T" handle or a double universal joint socket to remove the LEFT-HAND THREAD compressor wheel nut (2).**

Remove compressor wheel (3), then remove center housing from turbine shaft. Remove heat shield (19) and split ring seal (20) from turbine shaft. Straighten tabs of lockplates (27), remove cap screws (26), then remove center housing from back plate. Remove oil baffle (8) and seal ring (5) from back plate (4). Remove oil slinger (7) from back plate and split ring seal (6) from oil slinger. Remove Torx head screws (9), then remove thrust bearing (10) and thrust collar (11). Remove outer snap rings (14 and 17) from each end of center housing and, using a wire hook, pull bearings (15 and 18) from center housing. Remove inner snap rings (14 and 17) from housing, being careful not to damage housing bore.

Clean all parts in a suitable carbon softening solvent. A stiff fiber brush and plastic or wood scraper should be used after deposits have softened. When cleaning, use extreme caution to prevent parts from being nicked, scratched or bent.

Inspect turbine housing (22) outlet opening for warping or cracks. Check compressor housing (1) and back plate (4) for wheel friction wear. Inspect turbine shaft for grooves or scratches and turbine wheel and compressor wheel for bent or cracked blades or any other damage. Do not attempt to straighten bent turbine or compressor wheel blades. Check heat shield (19) and oil slinger (7) for damage and center housing (16) bearing surfaces for scoring or scratches. If any of these parts are damaged, renew turbocharger as they are not serviced separately.

A kit consisting of the following parts is available for cleaning and servicing the turbocharger: nut (2), seal ring (5), split ring seals (6 and 20), Torx head screws (9), thrust bearing (10), cap screws (12) with lockwashers, snap rings (14 and 17), bearings (15 and 18), lockplates (24 and 27), cap screws (25 and 26) and gasket (28).

Lubricate all parts with clean engine oil and reassemble by reversing the disassembly procedure. When installing snap rings (14 and 17), make sure round edge of snap rings face the bearings. When

installing nut (2), use a double universal joint socket and tighten to a torque of 14 N·m (120 in.-lbs.). Other tightening torques are as follows: center housing-to-back plate screws (26)—6 N·m (48 in.-lbs.); Torx head screws (9)—5 N·m (36 in.-lbs.); center housing-to-turbine housing cap screws (25)—11 N·m (96 in.-lbs.); back plate-to-compressor housing cap screws (12)—6 N·m (48 in.-lbs.).

With unit assembled, use a dial indicator and check turbine shaft horizontal end play, which should be 0.10-0.16 mm (0.004-0.006 in.). If shaft end play exceeds 0.16 mm (0.006 in.), thrust collar (11) and/or thrust bearing (10) is worn excessively. End play of less than 0.10 mm (0.004 in.) indicates incomplete cleaning (not all carbon removed) or dirty assembly and unit should be disassembled and recleaned.

Reinstall turbocharger, then prime turbocharger as outlined in paragraph 73.

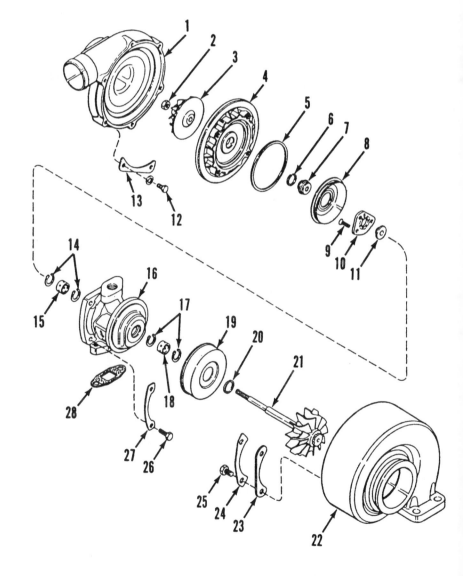

Fig. 70—Exploded view of typical turbocharger used on Models 5120 and 5140.

1. Compressor housing
2. Nut
3. Compressor wheel
4. Back plate
5. Seal ring
6. Split ring seal
7. Oil slinger
8. Oil baffle
9. Screw (3)
10. Thrust bearing
11. Thrust collar
12. Cap screw
13. Clamping plate
14. Snap rings
15. Bearing
16. Center housing
17. Snap rings
18. Bearing
19. Heat shield
20. Split ring seal
21. Turbine wheel & shaft
22. Turbine housing
23. Clamping plate
24. Lockplate
25. Cap screw
26. Cap screw
27. Lockplate
28. Gasket

AFTERCOOLER

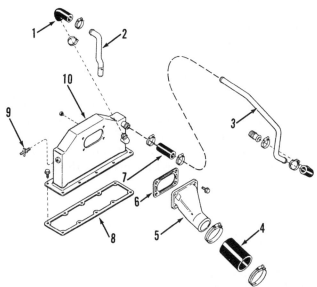

Fig. 71—View of aftercooler and relative parts used on Model 5120 prior to engine serial number JJF1005939.

1. Return hose
2. Return tube
3. Supply tube
4. Air hose
5. Air tube
6. Gasket
7. Supply hose
8. Gasket
9. Air bleed valve
10. Aftercooler

Model 5120 Prior to Engine S/N JJF1005939

75. REMOVE AND REINSTALL. To remove the aftercooler (10—Fig. 71), open bleed valve (9) and drain cooling system. Loosen clamps on air hose (4), then unbolt and remove air tube (5). Disconnect and remove the fuel injector lines. Immediately cap or plug all fuel openings. Disconnect coolant hoses (1 and 7), then unbolt and remove aftercooler (10).

NOTE: If any damage to aftercooler is evident, renew the complete aftercooler assembly.

Clean mounting surfaces of aftercooler and cylinder head of all foreign material. Using a new gasket (8), install aftercooler. Coat threads of cap screws with Loctite liquid Teflon, then install and tighten cap screws to a torque of 24 N·m (18 ft.-lbs.). Follow the tightening sequence shown in Fig. 31. Complete installation by reversing the removal procedure, keeping the following points in mind: Use new gasket (6—Fig. 71) on air tube (5) and tighten cap screws to a torque of 24 N·m (18 ft.-lbs.). When refilling with coolant, open air bleed valve (9) to remove air from aftercooler. When coolant appears at valve, close the valve and complete filling the system.

AIR CLEANER

All Models

76. FILTERS. All models are equipped with a dry-type air cleaner with a safety filter element (2—Fig. 72). This element should be renewed each 1000 hours of operation. DO NOT attempt to clean safety element.

Large primary filter element (4) can be cleaned by directing compressed air into inside of element. Air pressure must not exceed 690 kPa (100 psi). Keep air nozzle 125 mm (5 in.) from element and move nozzle up and down while turning the element. If element is covered with oil or soot, renew element.

After every 50 hours of operation, squeeze rubber dust valve (7) to remove dust accumulation.

Tractor is equipped with an air filter restriction switch and a warning lamp on the digital instrument cluster, which will indicate when primary element (4) needs service.

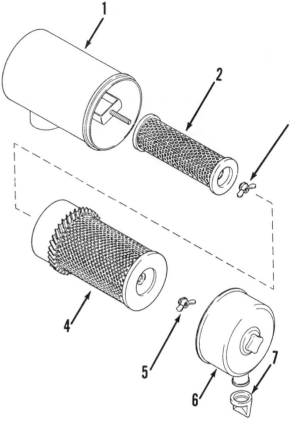

Fig. 72—Exploded view of air cleaner assembly used on all models.

1. Air cleaner housing
2. Safety element
3. Wing nut
4. Primary element
5. Wing nut
6. End cover
7. Dust unloader valve

COOLING SYSTEM

RADIATOR

All Models

77. REMOVE AND REINSTALL. To remove the radiator, drain cooling system and disconnect front wiring harness. Remove hood, side panels and grille. Disconnect and remove air intake duct. Unbolt and remove air cleaner and ether injector system. Unbolt and move air conditioning condenser, if so equipped, rearward and lay on engine. It is not necessary to disconnect the air conditioner lines from the condensor. Identify hydraulic oil cooler lines, disconnect lines and remove oil cooler. Disconnect upper and lower radiator hoses. Attach a hoist to radiator supports and unbolt supports from front support. Slide radiator assembly forward so that shroud will clear fan as the assembly is removed. Unbolt and remove upper air baffle, fan shroud and coolant recovery tank. Unbolt and remove radiator from supports.

Reinstall by reversing the removal procedure. Make certain that all foam rubber seals and pads are correctly installed in their original positions. Refill radiator with coolant. On Model 5120 equipped with aftercooler, open bleed valve on aftercooler to remove air trapped in aftercooler. Coolant capacity is 17.6 liters (18 U.S. qt.) for Model 5120 tractors and 20.8 liters (22 U.S. qt.) for Models 5130 and 5140. On Model 5120 with aftercooler, coolant capacity is 19.2 liters (20 U.S. qt.).

A 105 kPa (15 psi) radiator cap is used on all models.

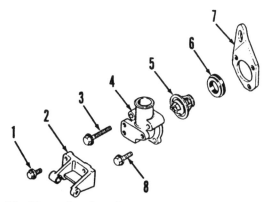

Fig. 73—View showing thermostat and relative components.

1. Cap screw (3)	5. Thermostat
2. Alternator bracket	6. Housing seal
3. Cap screw (2)	7. Front lift bracket
4. Thermostat housing	8. Cap screw

THERMOSTAT

All Models

78. REMOVE AND REINSTALL. To remove the thermostat (5—Fig. 73), first drain cooling system and proceed as follows: Raise belt tensioner and remove fan belt. Remove alternator, then unbolt and remove alternator mounting bracket (2). Remove upper radiator hose from thermostat housing. Remove cap screws (3 and 8) and lift off thermostat housing (4) with thermostat (5), seal (6) and engine front lift bracket (7). Remove lift bracket, seal and thermostat from housing. Before installation, test thermostat for correct temperature. Thermostat should start to open at approximately 83° C (181° F) and should be completely open at 95° C (203° F).

Install thermostat so that tang on thermostat is in slot of the housing. Install seal (small diameter first) in housing, then place lift bracket over the seal. Install the assembly and tighten cap screws to a torque of 24 N•m (18 ft.-lbs.). Install alternator bracket (2) and tighten cap screws (1) to 24 N•m (18 ft.-lbs.) torque. The balance of installation is the reverse of removal procedure.

WATER PUMP

All Models

79. REMOVE AND REINSTALL. To remove the water pump assembly (8 and 9—Fig. 74), first drain cooling system and proceed as follows: Raise belt tensioner (11) and remove fan belt (5). Remove retaining cap screws and withdraw water pump with "O" ring (10).

Water pump (9) and pulley (8) are available only as an assembly. When installing, use new "O" ring (10) and tighten water pump housing cap screws to a torque of 24 N•m (18 ft.-lbs.).

FAN

All Models

80. REMOVE AND REINSTALL. To remove fan (7—Fig. 74), raise the hood and remove right side panel. Raise belt tensioner (11) and remove fan belt (5). Unbolt fan and move forward into fan shroud. Remove spacer (6) and pulley (4). Lift fan (7) from shroud. Unbolt and remove fan bracket assembly (1, 2 and 3), if necessary.

Reinstall fan by reversing removal procedure. Tighten fan bracket bolts to a torque of 24 N•m (18 ft.-lbs.) and fan mounting bolts to a torque of 43 N·m (32 ft.-lbs.).

BELT TENSIONER

All Models

81. REMOVE AND REINSTALL. To remove the belt tensioner (11—Fig. 74), raise hood and remove

right side panel. Raise belt tensioner and remove fan belt. Remove mounting bolt and lift out belt tensioner. Remove Allen head bolts and remove tensioner mounting bracket (12), if necessary.

Reinstall bracket (12) and tighten bolts to a torque of 24 N·m (18 ft.-lbs.). Install tensioner (11) and tighten mounting bolt to a torque of 43 N·m (32 ft.-lbs.).

Before installing fan belt, refer to Fig. 75 and check belt tensioner as follows: Using a torque wrench, raise tensioner pulley about 45 degrees from perpendicular. Then, lower tensioner pulley to 35 degrees from perpendicular and check torque reading. Torque reading should be 24 N·m (18 ft.-lbs.). If torque reading is less then 24 N·m (18 ft.-lbs.), renew belt tensioner. Raise tensioner and install fan belt.

Fig. 74—View showing typical water pump, belt tensioner and fan with component parts. Items 1, 2 and 3, and items 8 and 9 are serviced only as assemblies.

1. Bracket
2. Shaft and bearing
3. Hub
4. Fan puller
5. Belt
6. Spacer
7. Fan
8. Water pump pulley
9. Water pump
10. "O" ring
11. Belt tensioner
12. Bracket

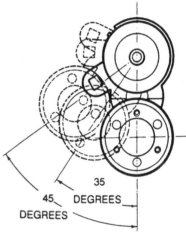

Fig. 75—View showing tension check of fan belt tensioner. Refer to text.

ELECTRICAL SYSTEM

BATTERIES

All Models

82. Before any electrical system service is performed, a thorough check of battery condition, condition of cable connections and condition of alternator drive belt should be made.

Batteries should be checked for total voltage and voltage drop under rated load. If battery voltage is excessively low, it should be recharged to rated level using an external battery charger. Failure to do so can cause alternator to overheat resulting in premature failure of alternator, regulator, or both.

All relative connections should be checked for excessive resistance, using an ohmmeter.

83. BATTERY CURRENT DRAIN TEST. If after checking, servicing and installing batteries, they return to an undercharged condition when tractor is not in use, current drain should be suspected. To check, disconnect negative ground cables at starter ground post and connect an ammeter lead to starter negative post and the other lead to negative cable ends. Make sure all electrical systems are off. Any reading on ammeter indicates current leakage. Check for lights, radio or accessories being on, or shorted electrical wiring or components.

CHARGING SYSTEM

All Models

84. TESTING. Prior to beginning test, be sure batteries are fully charged, all connections are clean and tight and alternator drive belt is in good condition and under correct tension.

> **CAUTION: Do not disconnect any wires from alternator while engine is running as damage to alternator may result.**

Alternator warning light should not come on when key switch is in OFF position. If light fails to go out when key is in OFF position, test circuits as outlined in paragraph 85.

Normal operation of alternator warning light is to come on only when key switch is in ACC or ON position, engine NOT running. If engine is started while throttle is in low idle position, light may stay on until engine speed is increased for the first time. Light should not come on at any time after engine is running and initial engine speed increase was sufficient to shut light off. If operation of warning light is not normal, test circuits as outlined in paragraph 85.

85. WARNING LIGHT CIRCUIT TEST. If warning light fails to go out with key switch in OFF position, disconnect wire from B+ terminal on alternator. If light goes out when B+ wire is disconnected, rectifier bridge in alternator is defective and must be renewed.

If light stays on after disconnecting wire from B+ terminal on alternator, there is an electrical short between wires at some place in wiring harness. Check wire insulation for damage.

If light fails to come on with the key switch in ACC or ON position, engine NOT running, first check condition of indicator bulb and bulb connection. If bulb tests satisfactory, check for a break in the circuit to the instrument cluster or an open circuit in the regulator or alternator. Check for the problem as follows: Test the voltage at key switch. If voltage is not approximately 12 volts, check for continuity from B+ terminal on alternator to the key switch. If this circuit is good, check batteries. If voltage at key switch is 12 volts, connect a 12-volt test light between D+ terminal on alternator and the key switch. If the test light comes on, the circuit between the indicator and the key switch or the circuit between the D+ terminal and the indicator is open. Check connections and wiring. If test light fails to come on, there is an open circuit in the alternator or voltage regulator.

If warning light functions normally with engine not running, but fails to go OFF when engine is running at various speeds, check alternator as outlined in paragraphs 86 and 87.

86. ALTERNATOR VOLTAGE OUTPUT TEST. Check to see that all accessories are turned OFF. With engine operating at rated speed (2200 rpm), connect the positive lead of a voltmeter to the B+ terminal of the alternator and the negative lead to a good ground on the engine. If voltmeter reading is less than 13 volts or in excess of 15 volts, disassemble alternator and check voltage regulator and other components.

87. ALTERNATOR AMPERAGE OUTPUT TEST. Connect an ammeter in the circuit at the B+ terminal on the alternator. See Fig. 76. Connect battery load tester (CAS-10147 or equivalent) across battery terminals. Turn light switch to ROAD position. With engine operating at rated speed (2200 rpm), adjust battery load tester for maximum current output. If reading is 65 amperes or above for alternator part number A187873 or 95 amperes or above for alternator part number A187623, and alternator warning light is still ON, disassemble alternator and test the rectifier bridge.

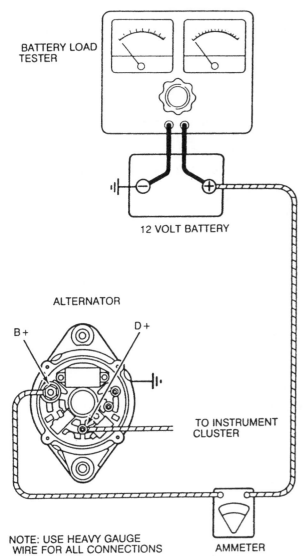

Fig. 76—Connect ammeter and battery load tester as
shown when checking amperage output of alternator.

ALTERNATOR

All Models

88. Robert Bosch 65 ampere (part number
A187873) without cab, or 95 ampere (part number
A187623) with cab, alternators are used on all models. Alternator is equipped with a solid-state regulator and has no provisions for adjustment.

Rated output is 14 volts at either 65 amperes or 95
amperes.

> CAUTION: Because certain components of the
> alternator can be damaged by procedures that will
> not affect a DC generator, the following precautions
> MUST be observed:

• Before working on the electrical system, disconnect the battery cables.

• When installing batteries, negative post of batteries must be grounded.

• When using an auxiliary battery for starting,
connect negative to negative and positive to positive.
Use the auxiliary battery connectors provided on
tractor.

• When charging batteries on tractor, use the auxiliary battery connectors on tractor. Do not attach
charger directly to battery terminals.

• Never operate tractor when battery cables are
disconnected.

• If necessary to weld on tractor, disconnect battery
cables and instrument cluster. Connect welder
ground cable as close as possible to weld area.

89. DISASSEMBLY AND TESTING. To disassemble the removed alternator, remove pulley nut
(20—Fig. 77), washer (19), tapered collar (18), pulley

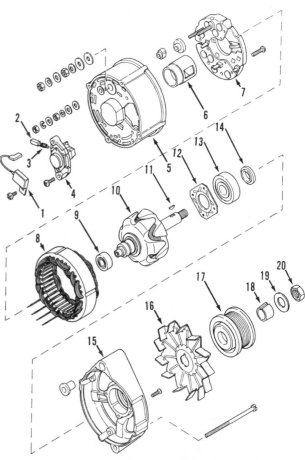

Fig. 77—Exploded view of typical 65-ampere or 95-am-
pere Robert Bosch alternator used on all models.

1. Capacitor	11. Key
2. Brush (2)	12. Plate
3. Brush spring (2)	13. Front bearing
4. Brush holder & voltage regulator	14. Spacer
5. Rear housing	15. Front housing
6. Shield	16. Fan
7. Rectifier	17. Pulley
8. Stator	18. Tapered collar
9. Rear bearing	19. Washer
10. Rotor	20. Nut

(17), fan (16) and key (11). Remove voltage regulator and brush holder (4) with springs (3) and brushes (2). Remove capacitor (1). Place alignment marks across front housing (15), stator (8) and rear housing (5) for aid in correct reassembly. Remove the four through-bolts and separate front housing and rotor assembly from rear housing and stator assembly. Remove the four screws securing retainer plate (12) to front housing (15), then separate rotor assembly from front housing. Using a suitable puller, remove rear bearing (9) from rotor, then press retainer plate (12), front bearing (13) and spacer (14) from front of rotor (10). Remove retaining screws and remove rectifier (7), stator (8) and shield (6) from rear housing (5).

Check the rotor for damage. Inspect slip rings for wear or deep grooves. Slip rings can be cleaned up using fine (400 grit) sandpaper. Touch one ohmmeter lead to the rotor frame and the other lead to a slip ring, as shown in Fig. 78. Then, repeat the test on the other slip ring. There must be no continuity between

either slip ring and rotor frame. A reading near zero will indicate a short circuit to ground. Touch ohmmeter leads to each slip ring as shown in Fig. 79. Ohmmeter reading should be 2.6-3.0 ohms. A higher reading will indicate an open circuit and a lower reading will indicate an internal short. If rotor fails to meet any test, renew rotor.

Measure the length of brushes. Renew brushes if the measurement is 14 mm (0.55 in.) or less. Brushes must work freely in holder.

To test the positive diodes in the rectifier bridge, proceed as follows: The positive diodes are the seven diodes closest to the B+ terminal. Connect the negative lead of the ohmmeter to the B+ terminal (Fig. 80). Connect the positive lead of the ohmmeter to one of the leads to the positive diodes and record the reading. Reverse the ohmmeter leads and check the ohmmeter reading. There should be one high reading and one low reading. If both readings are the same, renew rectifier bridge. Check the other six positive diodes in the same manner. To test the negative diodes, proceed as follows: The negative diodes are the seven diodes on the side of the rectifier bridge opposite the B+ terminal. Connect the negative ohmmeter lead to the unpainted surface of the diode plate. Connect the positive lead of ohmmeter to one of the leads to the negative diodes and record the reading. Reverse the ohmmeter leads and check the ohmmeter reading. There must be a high reading and a low reading. If both readings are the same, renew rectifier bridge. Repeat the test for the other six negative diodes.

Measure the resistance of the stator windings by connecting ohmmeter leads to stator leads. The resistance in each coil should be 0.1 ohm. If not, renew the stator. To check the stator for a short circuit, connect one lead of ohmmeter to the stator frame and the other lead to each of the stator leads. If the ohmmeter moves, there is a short circuit in the stator and stator must be renewed.

To test the voltage regulator, use voltage regulator tester (CAS-10851). Connect the red positive lead of

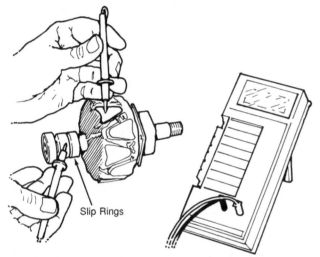

Fig. 78—Using an ohmmeter, check for ground between slip rings and rotor frame.

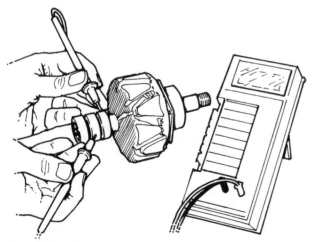

Fig. 79—Ohmmeter reading between slip rings should be 2.6-3.0 ohms.

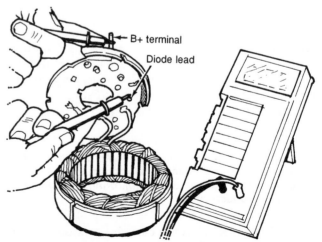

Fig. 80—Use multimeter to test diodes. Refer to text.

tester to the D+ brush, the yellow field lead to the DF brush and the black ground lead to the D– terminal. See Fig. 81. Depress test switch to the upper position. The green light must come on. Observe the voltage reading for 20 to 30 seconds. The voltage must be in the regulated range of 13.7 to 14.5 volts. If not, renew the voltage regulator and brush holder assembly.

90. REASSEMBLY. Reassemble the alternator by reversing the disassembly procedure, keeping the following points in mind: Use rosin-core solder when attaching stator leads to connections on rectifier bridge. It is recommended that bearings (9 and 13—Fig. 77) be renewed any time they have been removed. Tighten the four through-bolts evenly to a torque of 4-5 N•m (3-4 ft.-lbs.). Tighten pulley nut to a torque of 68 N•m (50 ft.-lbs.).

STARTING MOTOR

All Models

91. Robert Bosch starting motors are used on all models. Starting motor part number A186180 is used on Model 5120 tractors and starting motor part number A187597 is used on Model 5130 and Model 5140 tractors. Service procedures for both starting motors are similar and any differences will be noted.

92. DISASSEMBLY AND TESTING. Before disassembling the starting motor, first perform a no-load test to help determine the cause of any problems in operation.

Starter No. A186180—
No-load test at 27° C (80° F)
Volts 12
Current draw (max.) . . . . 85 amps
Armature speed (min.). . . . . 7000 rpm

Starter No. A187597—
No-load test at 27° C (80° F)
Volts 12
Current draw (max.) . . . . 170 amps
Armature speed (min.). . . . . 8000 rpm

Some problems and their possible causes are as follows:

1. Low armature speed and high current draw. Could be caused by:
   a. Tight, dirty or worn bearings
   b. Bent armature shaft
   c. Loose pole shoes contacting armature
   d. Short circuit in armature coil
   e. Damaged field coil

2. Armature does not rotate and current draw is high. Could be caused by:
   a. Field terminal contacting field frame
   b. Damaged field coil
   c. Worn bearings causing armature to contact pole shoes

3. Armature does not rotate and current draw is zero. Could be caused by:
   a. Open field circuit
   b. Open armature coil
   c. Brushes not contacting commutator bars

4. Low armature speed and low current draw. Could be caused by:
   a. Dirt or corrosion on connections
   b. Damaged wiring
   c. Dirty commutator
   d. Excessively worn brushes or faulty brush springs

5. High armature speed and high current draw. Could be caused by:
   a. Short circuit within field coil

To disassemble the starting motor, first place alignment marks across drive housing (7—Fig. 82), field frame (3) and commutator end cover (19). Remove screws and end cap (15), seal (16), "U" washer (17) and shims (18). Remove nuts and washers from studs (2) and remove commutator end cover (19). Disconnect the motor lead from the solenoid switch (1). Remove brush holder (21) and field frame (3) assembly. Bend back spring retainer tabs, remove springs and brushes, then separate brush holder from field frame. Remove screws and withdraw solenoid switch

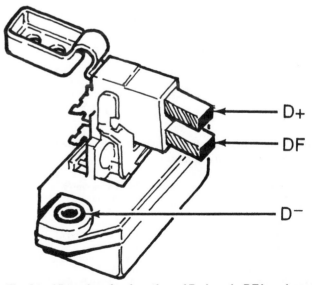

Fig. 81—View showing location of D+ brush, DF brush and D– terminal.

(1). Remove pivot bolt and remove armature assembly and shift lever (9) from front drive housing (7). Drive stop collar (11) on armature shaft, remove snap ring (10) and collar, then slide drive clutch (13) and center bearing (5) from armature (14).

Measure length of brushes. If brushes are worn to less than 8.0 mm ($\frac{5}{16}$ in.), install new brush kit (23).

Use a multimeter to check for continuity between commutator segment and armature core. If there is continuity, armature has a ground and must be renewed.

Place leads of the multimeter on two segments of the commutator. If there is no continuity, the winding has an open circuit and must be renewed.

With armature resting in "V" blocks, check commutator for out-of-round or run out. If out-of-round or run out exceeds 0.03 mm (0.001 in.), place armature in a lathe and true up the commutator. After truing the commutator, undercut the mica between segments to a depth of 0.4 mm ($\frac{1}{64}$ in.).

Check for continuity between field terminal and field frame. If there is continuity, field coil is grounded and must be renewed.

Check for continuity between field terminal and brush connector. If there is no continuity, field coil has an open and must be renewed.

Inspect drive clutch (13) for excessive wear or other damage and renew as necessary. Bushing (12) is renewable in drive clutch.

Inspect bushings (8 and 20) and center bearing (5) and renew as required.

93. REASSEMBLY. Reassemble starting motor in reverse order of disassembly, keeping the following points in mind: Use a new "O" ring (6—Fig. 82) and align marks during assembly. Lubricate bushings and drive clutch splines with GE silicone grease (G321 Versilube) or equivalent. Tighten nuts on studs (2) evenly to a torque of 4.5-6.0 N•m (40-53 in.-lbs.). Install shims (18) as required to provide armature end play of 0.05-0.4 mm (0.002-0.015 in.) on starting motor for Model 5120 or 0.1-0.3 mm (0.004-0.012 in.) on starting motor for Model 5130 or Model 5140. Use new seal (16) ιd install end cap (15).

WIRING

All Models

94. When a problem is suspected in the wiring harness, follow an orderly, step-by-step check of the problem circuit. Always check protective circuits (fuses and circuit breakers) first, then locate primary point of current supply for circuit in question. Using a 12-volt test light or a voltmeter, check for adequate voltage at this point. If voltage is present at this point, continue to check for voltage at appropriate connections along harness, working toward malfunctioning unit until problem area is isolated. Harness connectors (plugs), sockets and connection terminals being corroded, loose or broken are most often the cause for interruptions in power supply.

If voltage was not present at primary point of supply, check wiring moving toward battery until the current interruption is isolated. Repair as necessary.

PROTECTIVE CIRCUITS

All Models

95. FUSES. Fuses are installed in circuits to protect wiring from damage in event of shorted wiring.

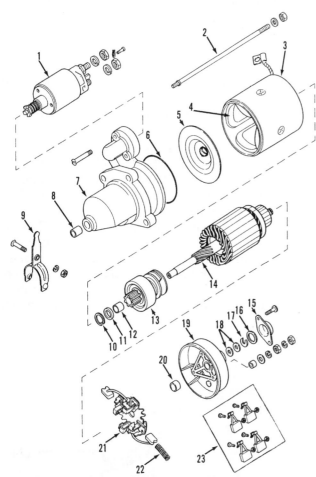

Fig. 82—Exploded view of typical Robert Bosch starting motor used on all models.

1.	Solenoid switch	13.	Drive clutch
2.	Stud (2)	14.	Armature
3.	Field frame	15.	End cap
4.	Field coils	16.	Seal
5.	Center bearing	17.	"U" washer
6.	"O" ring	18.	Shims
7.	Drive housing	19.	Commutator end cover
8.	Bushing	20.	Bushing
9.	Shift lever	21.	Brush holder
10.	Snap ring	22.	Brush spring
11.	Stop collar	23.	Carbon brush service kit
12.	Bushing		

When fuse is blown (fuse strip melted), a fuse of the same size and amperage must be installed after locating and repairing the problem. Fuses and circuit breakers are accessible after removing side panels at each side of instrument panel support on cab models or after opening the small access door at each side on ROPS models.

96. CIRCUIT BREAKERS. Circuit breakers perform the same function as fuses. However, a circuit breaker will cut off current flow, then cool down and reset automatically, re-establishing current flow. Usually used in lighting, cab circuits and gauge circuits, it eliminates complete failure allowing limited use of systems considered necessary for safe operation.

Circuit breakers will be in line with red wire connected to positive cable connection. Under normal operation with voltage at BAT connection of circuit breaker, voltage should pass through breaker to remaining terminal.

LIGHTS

All Models

97. Front and rear work lights, tail lights and warning lights are standard equipment. However, additional front and rear work lights are available.

Single light failure is usually a bad light bulb, poor ground or bad current connection. Multiple light failure is usually caused by bad switch or power source wiring problems. Check fuses and circuit breakers, then check wiring as outlined in paragraph 94.

INSTRUMENT CLUSTER

All Models So Equipped

98. STANDARD INSTRUMENT CLUSTER. Tractors may be equipped with the standard instrument cluster (Fig. 83).

The indicator lights (1) illuminate to alert the operator that there is a fault in a system or that the function chosen is operating. Refer to paragraph 100 for details.

The hourmeter (2) records engine running hours (white numbers) and tenths of hours (red number).

The tachometer/speedometer (3) is mechanically driven by a cable from the fuel injection pump drive gear nut. The upper part of the unit shows the engine speed (rpm). A symbol indicates the rpm for pto speed. The lower part shows tractor travel speed for each gear.

The fuel gauge (4) indicates the amount of fuel in the tank. The lower symbol indicates empty and upper symbol indicates full.

The engine coolant temperature gauge (5) is divided into three colored sections. The pointer in the yellow area indicates low temperature. The pointer in the green area indicates normal working temperature. The pointer in the red area indicates overheating. Stop engine and check the cause.

99. DELUXE DIGITAL INSTRUMENT CLUSTER. Tractors may be equipped with the digital instrument cluster (Fig. 84).

The indicator lights (1) illuminate to alert the operator that there is a fault in a system or that the function chosen is operating. Refer to paragraph 100 for details.

The pto rpm button (2), when pressed, displays the power take-off speed on the upper screen (5). The pto rpm is calculated from the actual engine speed. The

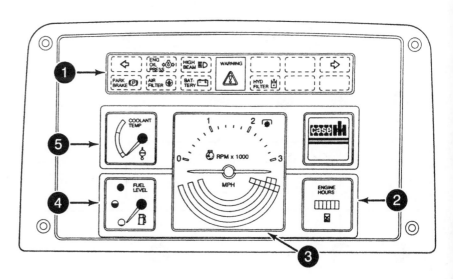

Fig. 83—View showing location of standard instrument cluster components.

1. Indicator lights
2. Hourmeter
3. Tachometer-speedometer
4. Fuel gauge
5. Engine coolant temperature gauge

pto speed will continue to display until the engine rpm button (3) is pressed.

The engine rpm button (3), when pressed, displays the engine speed on the upper screen (5). The engine rpm will continue to be displayed until the pto rpm button is pressed. The engine rpm will automatically be displayed when engine is started.

The hourmeter (4) records engine running hours (white numbers) and tenths of hours (red number).

The engine or pto rpm display screen shows the rpm depending upon which button (2 or 3) is pressed.

The tractor ground speed display (6) shows tractor travel speed in mph or km/h. Refer to programming instrument cluster in paragraph 102.

The wheel-slip indicator (7) shows the amount of wheel-slip as a percentage, each arrow indicates two percent. This operates only on a tractor equipped with a True Ground Speed (radar) unit.

The check mark (8) displays when key switch is turned ON. The check mark must go off when engine is started. If mark stays on when engine is started, engine rpm is not displayed, there is an open circuit on Wheel Speed Sensor or there is an open circuit on True Ground Speed Sensor. The mark will not be displayed if a fault occurs in these three circuits during operation, but will display when you stop and start the engine again.

The fuel gauge (9) shows the amount of fuel in the tank. Lower symbol is empty, center symbol is half-full and the upper symbol is full.

The engine coolant temperature gauge (10) is divided into three colored sections. The pointer in the yellow area indicates low temperature. The pointer in the green area indicates normal working temperature. The pointer in the red area indicates overheating. Stop engine and check the cause.

100. INDICATOR LIGHTS. The indicator lights shown in Figs. 85 and 86 are used on all models.

The green left-turn indicator light (1—Fig. 85) flashes on and off when indicating a left turn.

The red engine oil pressure light will illuminate when key switch is turned to the ON or START positions or if the oil pressure drops below the normal operating range during operation. If the light comes ON at any time when engine is running, stop the engine and check the cause.

The blue headlight high-beam indicator light (3) will be ON when light switch is turned to position 3, 4 or 5 and headlights are on high-beam. Headlights are switched to high- or low-beam by pulling end of turn signal lever upward (toward steering wheel).

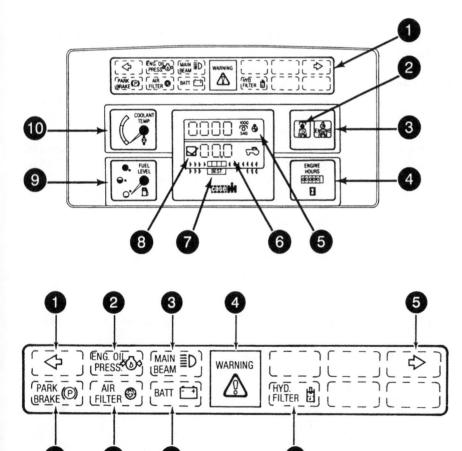

Fig. 84—View showing location of deluxe digital instrument cluster components.

1. Indicator lights
2. Pto rpm button
3. Engine rpm button
4. Hourmeter
5. Engine or pto rpm display
6. Tractor ground speed display
7. Wheel slip indicator
8. Check mark display
9. Fuel gauge
10. Engine coolant temperature gauge

Fig. 85—Identification of indicator lights used on all models.

1. Green left-turn indicator
2. Red engine oil pressure
3. Blue headlight high beam
4. Orange warning light
5. Green right-turn indicator
6. Red park brake
7. Orange air cleaner restriction
8. Red alternator charge light
9. Yellow hydraulic filter restriction

The orange warning light (4) will flash if reverse gear is selected when in fourth range gear and when the park brake is engaged.

The green right-turn indicator light (5) flashes on and off when indicating a right turn.

The red park brake light (6) illuminates when the park brake is engaged. An audible warning also sounds if a gear is engaged with park brake engaged.

The orange air cleaner restriction light (7) comes on if air cleaner becomes restricted. The primary element must be cleaned or renewed before the next period of operation.

The red alternator charge light (8) should come on when the key switch is turned to ON or START positions or if the alternator stops charging. If the light comes on during operation, stop the engine and check for cause.

The yellow hydraulic filter restriction light (9) illuminates if the hydraulic oil filter becomes restricted. Renew filter element no later than at the end of the day's work.

The amber powershift indicator light (10—Fig. 86) will come on if there is a fault in the powershift transmission system.

The green reverse selected indicator light (11) illuminates when reverse is selected.

The green differential lock indicator light (12) comes on if the differential lock is engaged.

The amber clutch disengage indicator light (13) will come on when key switch is turned to ON or ACC position or when a wrong selection has been made, such as engaging reverse when in fourth range.

101. CONTROL SWITCHES. Control switches located on lower instrument panel (Fig. 87) are as follows:

FRONT PTO SWITCH—Press front of switch (1) to operate the front pto (if so equipped) and press rear of switch to stop the unit.

FRONT-WHEEL DRIVE—To engage the front drive axle, press front of switch (2). A green light in the switch will come on in the engaged position. Press rear of switch to disengage.

KEY SWITCH—The key switch (3) is a four-position switch. Key can be removed only in the vertical OFF position. Use the ACCESSORY position to operate the radio when engine is not running. The engine oil pressure and alternator charge lights come on and the gauge indicators will show correct values when key is in ON position. After engine is started, lights should go OFF and engine rpm should be displayed on instrument cluster. This position is also used for programming the digital instrument cluster. When turned to the START position, park brake, master warning, hydraulic filter and engine air filter lights will be illuminated. Gauge indicators will be OFF and starter motor will crank engine. Release key switch when engine starts.

LIGHT SWITCH—The light switch (4) is a five-position switch. OFF is in horizontal position. Turn switch clockwise to the following positions: In first position, front and rear amber warning lights will flash. In second position, front and rear warning lights, tail lights, side console light and headlights will be ON. In third position, headlights, rear fender work lights and side console light will be ON. In fourth position, headlights, rear upper work lights, front upper work lights, rear fender work lights and side console light will be ON.

ETHER STARTING-AID SWITCH—When starting engine in cold temperatures of –6.7° C (20° F) or

Fig. 86—View of indicator lights on all lower instrument panels.

10. Amber check powershift
11. Green reverse selected indicator
12. Green differential lock
13. Amber clutch disengage indicator

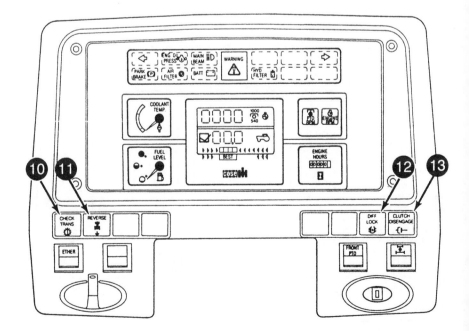

lower, press and hold front of switch (5) down for three seconds while cranking engine.

PROGRAMMING DIGITAL INSTRUMENT CLUSTER

All Models So Equipped

102. The digital instrument cluster must be programmed with variable data in order to calculate output information, such as ground speed and wheel slip. The digital cluster can be programmed to display readout information in U.S. customary or metric units of measurement.

To check the program code number, turn key switch to ON position. The cluster will perform a self-test for one second showing all digits. This is immediately followed by the program code number displayed on the upper screen (5—Fig. 84). The code number must be 004 for U.S. customary programmed cluster or 005 for metric programmed cluster. To change the program, turn key switch OFF. Press and hold both engine rpm and pto rpm buttons at the same time. While holding both buttons, turn key switch ON. The units will change automatically from one value to the other. Check to see that the correct program code is displayed.

The True Ground Speed Sensor is an option. If tractor is not equipped with this radar sensor, then ground speed is monitored by the transmission magnetic sensor. The True Ground Speed Sensor is located on the side of the transmission housing. The sensor must be mounted at the correct 35° angle. Use a carpenter's level to set mounting bracket lower edge at horizontal. Keep the face of the radar sensor clean. Scrape mud off with a plastic scraper. DO NOT use a metal scraper. Wash face of sensor with soap and water or steam clean. DO NOT disconnect wiring harness from sensor while cleaning.

A ground speed calibration code number must be programmed into the digital cluster in order to calculate ground speed accurately. There are three types of code numbers that can be used depending on how your tractor is equipped and how much accuracy is desired. Different types of code numbers are as follows:

Standard calibration code number 742 can be used only on tractors equipped with True Ground Speed Sensor. The standard code number is an average number representing the normal mounting angle and reading distance of the True Ground Speed Sensor.

The dynamic calibration is the most accurate and recommended method and is recommended for tractors with or without True Ground Speed Sensor.

The static calibration is the second most accurate method and can be used for tractors with or without True Ground Speed Sensor.

To program the standard code number into the digital cluster, make sure key switch is in OFF position. Remove retaining screws and remove cluster from the instrument panel. DO NOT disconnect wiring harness. Remove the cover over the CAL/OP switch, which is located at lower center on the back of the cluster. Turn key switch to ON position and slide CAL/OP switch to CAL. Observe the display screens on the cluster. The upper screen must show CAL, and the lower screen 742, which is the standard code entered at the factory. If the number is not correct, press the engine rpm button to increase the number or press the pto rpm button to decrease the number. Release the button when number 742 is displayed. Slide CAL/OP switch to OP, then turn key

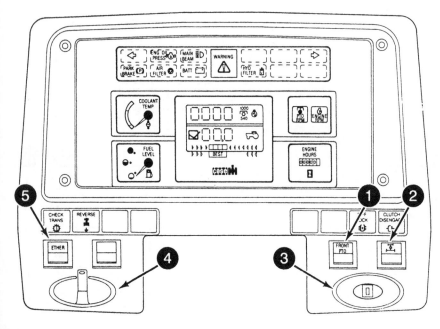

Fig. 87—Location of control switches on lower instrument panel.

1. Front pto
2. Front wheel drive
3. Key switch
4. Light switch
5. Ether starting aid

switch to OFF. Install cover over CAL/OP switch, then install the cluster and retaining screws.

To determine the static calibration code number, measure distance between center of rear axle (1—Fig. 89) and bottom of tire (2). Use this measurement and select the static calibration code number from the chart (Fig. 89). The procedure for entering the static calibration code number is the same as the standard code number.

To determine the dynamic calibration code number, measure a straight line course 60 meters (197 ft.) long. See Fig. 90. For best accuracy, choose a course surface of low grass crop or uniformly tilled ground. The next best surface, in order, is gravel, blacktop or concrete. Remove the digital cluster from the instrument panel. DO NOT disconnect the wiring harness. Remove cover over the CAL/OP switch and slide switch to CAL. Start driving the tractor toward the measured course. Keep the implement out of the ground to create a no-slip condition. As you pass the start mark, momentarily press either the engine rpm or the pto rpm button, then release. The display will reset to zero and begin to accumulate. As you pass the finish mark, push in, then release the engine pto or pto rpm button. The display will stop accumulating and will show the dynamic calibration code number on the upper screen and the radar sensor true ground speed code on the lower screen.

NOTE: If tractor is not equipped with True Ground Speed Sensor, display will show the code number on the upper screen and CAL on the lower screen.

Stop the tractor and slide the CAL/OP switch to OP. Install the switch cover, then install the digital cluster and retaining screws. The code number displayed will automatically be entered in the cluster memory when the switch is moved to OP position.

Tractor must be equipped with True Ground Speed Sensor in order to calculate and display wheel slip. The amount of wheel slip is indicated by a bar graph displayed across the bottom of the lower display screen. See Fig. 91. The instrument cluster compares the transmission magnetic sensor signal with the True Ground Speed Sensor signal. After wheel slip has been calibrated, wheel slip will be indicated anytime the tractor is in operation. The wheel slip bar

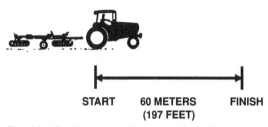

| | START | 60 METERS (197 FEET) | FINISH |

Fig. 90—To determine the dynamic calibration code number, tractor must be driven along a 60 m (197-ft.) straight-line course. Refer to text.

Fig. 89—Chart showing method of selecting static calibration code number.

STATIC CALIBRATION CODE NUMBERS			
STATIC LOADED RADIUS		STATIC CALIBRATION CODE NUMBER	MEASURING STATIC LOADED RADIUS
INCHES	MM		
24.5	620	635	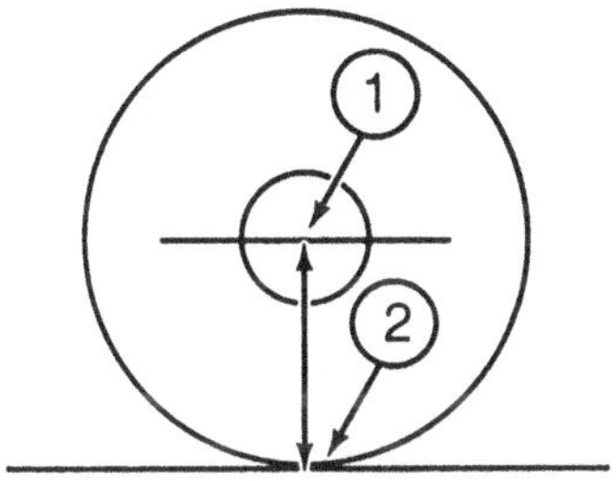
26.5	670	588	
27.0	690	571	
27.5	695	567	
28.0	715	551	
28.5	720	547	
29.0	740	532	
29.5	745	529	
30.0	765	515	
30.5	770	512	
31.0	775	508	
31.5	795	495	
32.5	820	480	
34.0	861	457	
36.5	930	423	

MEASURE FROM CENTER OF AXLE (1) TO BOTTOM OF THE TIRE (2).

Fig. 91—Wheel slip indicator used on tractors equipped with digital instrument cluster and True Ground Speed Sensor.

graph contains 15 segments. Each segment represents 2 percent slip. The range of the graph is 0 percent slip (all segments OFF) to 30 percent slip (all segments ON). The word BEST on the decal is aligned with segments 5, 6, 7 and 8 of the bar graph display. These segments represent a wheel-slip range of about 9 percent to 16 percent. Ideal wheel slip is 10 to 15 percent for drawbar work and 13 to 15 percent for hitch work.

TORQUE LIMITER CLUTCH

All models are equipped with a "slip-type" torque limiter clutch bolted to the engine flywheel. Both the Powershift and the Synchromesh speed transmissions receive power through an input shaft from the torque limiter clutch.

All Models

103. R&R AND OVERHAUL. To remove the torque limiter clutch, first split tractor between engine and speed transmission as follows: Apply park brake and securely block rear wheels. Disconnect battery cables and remove batteries and battery box. Unbolt and remove left step assembly. Drain fuel tank, disconnect breather hose, return hose and fuel supply hose, then unbolt and remove fuel tank. If so equipped, remove front drive shaft and shield as outlined in paragraph 9. Remove plugs and drain oil

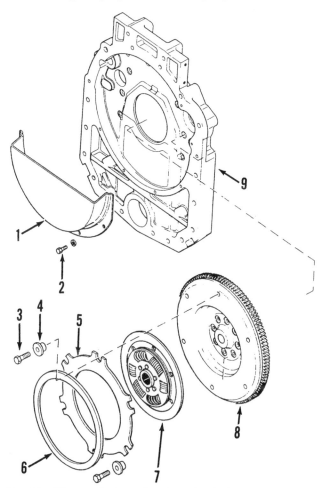

Fig. 92—Torque limiter clutch and relative components used on all models.

1. Oil baffle
2. Bolt
3. Bolt (6)
4. Spacer (6)
5. Steel plate
6. Belleville spring
7. Friction plate
8. Flywheel
9. Flywheel housing

from transmission and flywheel housings. Place wooden wedges between front axle and front support to prevent tipping. Raise the hood and remove exhaust extension pipe and rear hood cover. Disconnect hood release cable (if used) and move cable out of the way. Disconnect right steering tube and the supply and return oil cooler tubes. Disconnect wires and cable from starter motor. Close heater shut-off valve, then disconnect and drain heater hoses. Remove heater supply and return tubes. Disconnect throttle cable from fuel injection pump. Disconnect left steering tube. Loosen clamp and remove brake return hose from housing adapter. Disconnect engine wiring harness connector from fire wall. Disconnect ground cable and relay cables. If so equipped, disconnect air conditioning hoses at quick couplers. Loosen rear cab or platform mounting nuts and remove front cab or platform mounting bolts. Carefully tilt front of cab or platform and install blocks between cab or platform and mounting frame. If so equipped, remove front-end weights. Attach front split stand (CAS-10852) to support front of tractor and block up under front of speed transmission housing. Remove the four upper transmission-to-engine mounting bolts and the left and right outer side-rail bolts. Remove remaining transmission-to-engine bolts and carefully separate engine from transmission. Refer to Fig. 92 and unbolt and remove oil baffle (1) from flywheel housing (9). Remove bolts (3) and spacers (4), then remove Belleville spring (6), steel plate (5) and friction plate (7) from flywheel (8).

Clean and inspect all parts for wear or other damage. Measure free height of Belleville spring as shown in Fig. 93. Free height should be 10.82-11.33 mm (0.426-0.446 in.) for 25 mm (0.985 in.) wide spring or 11.27-11.78 mm (0.444-0.464 in.) for 40 mm (1.575 in.) wide spring.

When reassembling, install friction plate (7—Fig. 92) so that long offset side of hub is away from flywheel. Use special tool (CAS-1991) to center fric-

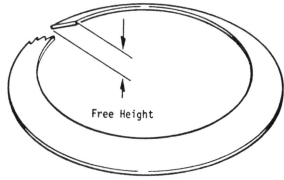

Fig. 93—Measure free height of Belleville spring as shown.

Free Height

tion plate. Install Belleville spring (6) so that raised outer diameter contacts flange of spacers (4). Tighten retaining bolts (3) to a torque of 134-151 N·m (100-110 ft.-lbs.).

Clean engine and transmission mounting flanges and apply a bead of Loctite 515 to engine mounting flange.

The balance of installation is the reverse order of removal. Tighten 12 mm engine to transmission bolts to 134-151 N·m (100-110 ft.-lbs.) and 16 mm engine to transmission bolts to 335-375 N·m (248-277 ft.-lbs.). Tighten cab or platform mounting bolts to 312-380 N·m (230-280 ft.-lbs.).

POWERSHIFT
TRANSMISSION

All models are available with a 4-speed Powershift transmission. This transmission, in conjunction with the 4-speed range transmission and the forward-reverse shuttle transmission, provides 16 forward and 12 reverse speeds. Fourth range is not used in reverse. When optional creeper drive is used, 8 low speeds forward and 8 speeds reverse are provided as range 3 and 4 are locked out.

OPERATION

All Models So Equipped

104. The Powershift is the system used to change gears 1 to 4 without needing to cycle the inching pedal. The movements of the Powershift transmission lever are converted into electrical signals at the shifter assembly (Fig. 94) and transmitted to the Powershift control module. The Powershift control module receives electrical signals from the shifter assembly and from the pressure switches in the Powershift manifold (Fig. 95). Depending upon signals received, the control module operates the solenoids in the manifold. The solenoids operate the hydraulic circuits that control the clutches of gears 1 to 4.

A fault in the Powershift system will be indicated by the amber Powershift indicator light on the instrument panel. The indicator light should come ON when key switch is turned to ON position. If there is no fault in the system, the light will go out when

engine is started, a gear is selected and the inching pedal is fully cycled. If there is a fault in the Powershift system, indicator light will remain ON or will start flashing.

Operation of tractors equipped with Powershift transmission is controlled by three levers as shown in Fig. 96: forward-reverse, Powershift and range. When selecting speeds with tractor stationary, select forward or reverse, move Powershift lever to position 1, fully depress clutch (inching) pedal and select range gear. Slowly release clutch (inching) pedal. Tractor will start to move. When selecting speeds with tractor moving, Powershift lever can be moved to any position without depressing the inching pedal. For smooth speed changes, move control lever one speed at a time. Always downshift one speed at a time. When changing direction of travel, always slow tractor speed before moving the control lever to the desired direction of travel. Neutral is provided in the range transmission. Always depress the inching pedal when shifting the range transmission.

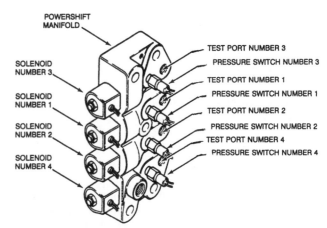

Fig. 95—Powershift manifold located on right side of transmission.

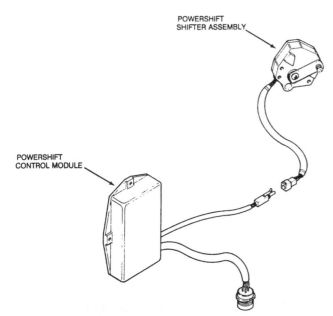

Fig. 94—Powershift shifter assembly and control module located in the right side console.

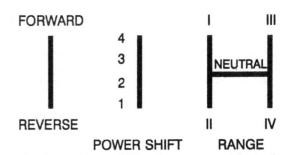

Fig. 96—Control lever shift patterns on models equipped with Powershift transmission.

TROUBLESHOOTING

All Models So Equipped

105. If the Powershift transmission fails to shift properly, problems may be either hydraulic or electrical. If the amber Powershift indicator light (10—Fig. 86) comes ON during tractor operation, fully depress inching pedal for two seconds. The indicator light will flash. Count the number of flashes and check for electrical problems as follows:

ONE FLASH—The Powershift control module is receiving either no electrical signals or multiple electrical signals from Powershift shifter (Fig. 94). Could be caused by:
  1. Wires shorted or faulty connections between control module and shifter assembly
  2. Faulty shifter assembly

TWO FLASHES—Off-going pressure at the Powershift manifold not detected. Could be caused by:
  1. Open in pressure switch circuit
  2. Faulty pressure switch (stuck open)
  3. Solenoid stuck in ON position

THREE FLASHES—The hydraulic oil temperature sensor circuit is open or shorted to ground. Could be caused by:
  1. Faulty wiring or connections between temperature sensor and control module
  2. Faulty temperature sensor
  3. Faulty control module

FOUR FLASHES—Solenoid circuit is shorted to ground. Could be caused by:
  1. Faulty wiring or connections between solenoid and control module
  2. Faulty solenoid

FIVE FLASHES—On-coming pressure at the Powershift manifold not detected. Could be caused by:
  1. Faulty pressure switch (stuck closed)
  2. Pressure switch circuit shorted to ground
  3. Open in solenoid circuit
  4. Solenoid stuck in OFF position

SIX FLASHES—No clutch pressure or multiple clutch pressures. Could be caused by:
  1. Pressure switch circuit shorted to ground
  2. Faulty pressure switch (stuck closed)
  3. Open circuit in one or more pressure switch circuits
  4. One or more solenoids stuck in ON position
  5. Faulty control module

Before testing for hydraulic faults, check the regulated pressure as follows: Remove the plug from test port on top of hydraulic pump compensator. Install fitting (CAS-2009-4) and a 4000 kPa (600 psi) test gauge. Operate engine until the hydraulic oil is heated to a temperature of 49° C (120° F). With engine speed at 1500 rpm, fully cycle the inching pedal. At this time, the regulated pressure must be 1800-2070 kPa (260-300 psi). Record the pressure reading and proceed as follows: Remove the plugs from the four test points on Powershift manifold (Fig. 95). Install four 4000 kPa (600 psi) pressure gauges, using extension tubes (CAS-2009-8). With hydraulic oil heated to 49° C (120° F) and engine operating at 1500 rpm, fully cycle inching pedal. Shift to each of the four gears and record gauge pressures. The pressure readings must be within 69 kPa (10 psi) of the previously recorded regulated pressure. Also, the differences between pressure readings for each gear must not vary more then 69 kPa (10 psi).

If pressure readings are not correct, check for the following possible causes:
  1. Faulty clutch pack
  2. Powershift spool damaged
  3. Internal leakage in tubes, seals or shafts
  4. Powershift manifold inlet screen plugged

POWERSHIFT MANIFOLD (CONTROL VALVE)

All Models So Equipped

106. R&R AND OVERHAUL. To remove the Powershift control valve (manifold), first disconnect battery cables, then remove batteries and battery box. Disconnect Powershift supply tube. Place identification tags on all electrical connectors to aid in correct reassembly. Disconnect the connectors and remove the ground wire. Remove the Powershift valve retaining bolts and lift off the control valve and gasket.

To disassemble the unit, remove nut (11—Fig. 97) and first "O" ring. Remove solenoid coil (10) and second "O" ring. Unscrew and remove valve spool (9) with "O" rings (8). Remove the other three solenoid valves in the same manner. Unscrew and remove the four pressure switches (5). Remove test port plugs (4) and inlet elbow (7) with "O" rings (3 and 6).

Clean and inspect all parts for damage. Using an ohmmeter, check resistance of solenoid coil. Reading should be 8 ohms. Check for continuity between pressure switch leads. The normally closed switch should show continuity. Solenoid coils (10), valve spools (9) and pressure switches (5) are not serviceable; renew faulty or damaged parts.

Using all new "O" rings and gasket (1), reassemble and reinstall by reversing the disassembly and removal procedures.

HYDRAULIC OIL
TEMPERATURE SENSOR

All Models

107. The hydraulic oil temperature sensor is located in the hydraulic oil filter housing. Removal of the sensor is obvious after examination of the unit. Using an ohmmeter, check resistance of sensor against the temperatures shown in chart in Fig. 98. Renew sensor if it fails to meet the specifications.

POWERSHIFT UNIT

All Models So Equipped

108. R&R AND OVERHAUL. To remove the Powershift transmission, first split tractor between the engine and transmission housing as outlined in paragraph 103.

Disconnect lubrication lines and clutch supply lines from input shaft and drop shaft front bearing covers as shown in Fig. 99. Unbolt and remove access cover (just rearward from Powershift valve), then remove bolt, washer, any shims present and drive gear (Fig. 100) from rear of drop shaft. Unbolt and remove input shaft and drop shaft front bearing covers (1 and 2—Fig. 99) and shims. Remove the four retaining bolts, then lift out front bearing carrier plate (Fig. 101). Remove the input (top) shaft assem-

bly and place on a clean bench. Lift out drop (lower) shaft assembly and place on bench.

To disassemble the input shaft, remove seal rings (43—Fig. 102) and bearing cone (42). Remove snap ring (13) from its groove in second gear clutch housing (3). Remove bearing cone (1), "O" ring (2) and second gear clutch housing (3). Place input shaft assembly in a press and, using special tool (CAS-1992) against piston (7), compress piston return spring (14). Remove snap ring (4), then release press. Remove piston

HYDRAULIC OIL TEMPERATURE		RESISTANCE (Ohms)
°C	°F	(±15%)
-17.8	0.0	1787
-12.2	10.0	1535
-6.7	20.0	1264
-3.9	25.0	1154
-1.1	30.0	1073
0.0	32.0	1066
4.4	40.0	905
10.0	50.0	790
15.6	60.0	677
21.1	70.0	595
26.7	80.0	521
32.2	90.0	444
37.8	100.0	390
43.3	110.0	344
48.9	120.0	298
54.4	130.0	265
65.5	150.0	209
76.7	170.0	166
87.8	190.0	135
98.9	210.0	112

Fig. 98—Chart used when checking hydraulic oil temperature sensor.

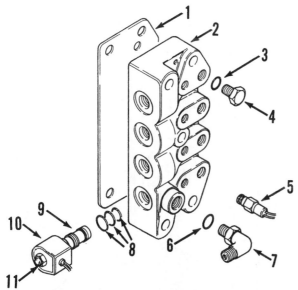

Fig. 97—Exploded view of Powershift control valve.

1. Gasket
2. Manifold
3. "O" ring
4. Test port plug (4)
5. Pressure switch (4)
6. "O" ring
7. Inlet elbow
8. "O" rings
9. Valve spool (4)
10. Solenoid coil (4)
11. Nut

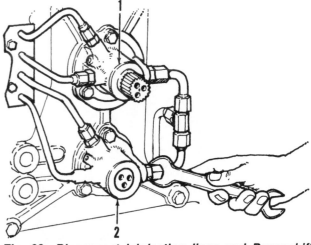

Fig. 99—Disconnect lubrication lines and Powershift clutch lines from input shaft front bearing cover (1) and drop shaft front bearing cover (2) as shown.

(7) with seal rings (5 and 8) and "O" rings (6 and 9). Remove separator plates (10), friction plates (11), backing plate (12), snap ring (13), piston return spring (14), spring retainer (15), snap ring (16), thrust washer (17) and thrust bearing (18). Remove first gear (19), roller bearings (20), spacer (21), thrust washer (22), thrust bearing (23), third gear (24), roller bearings (25), spacer (26), thrust bearing (27) and thrust washer (28). Using special tool (CAS-1992) and a press, compress piston return spring (31) and remove snap ring (29). Release press and remove spring retainer (30), spring (31), snap ring (32), backing plate (33), friction plates (34), separator plates (35) and piston (38) with seal rings (36 and 40) and "O" rings (37 and 39) from input shaft (41).

Clean and inspect all parts and renew any showing excessive wear or other damage. Thickness of new friction plates (11 and 34) is 2.45-2.60 mm (0.096-0.102 in.). Thickness of new separator plates (10 and 35) is 2.11-2.31 mm (0.083-0.091 in.).

Use Figs. 102 and 104 as guides and reassemble input shaft assembly in reverse order of disassembly, keeping the following points in mind: Use all new "O" rings and seal rings during reassembly. Soak friction plates (11 and 34—Fig. 102) in clean Hy-Tran Plus oil before installation. Heat bearing cones (1 and 42) in a bearing oven to a temperature of 150° C (302° F) when installing. Install seal rings (43) on input shaft (41) using seal expander (CAS-2002), then compress seal rings into grooves in input shaft using seal compressor (CAS-2005-4).

To disassemble the drop shaft assembly, refer to Fig. 103 and remove seal rings (41), bearing cone (40), thrust washer (39), thrust bearing (38), fourth gear (36), roller bearings (37), thrust bearing (35) and thrust washer (34). Use a press and spring compressor tool (CAS-1992) to compress piston return spring (31) and remove snap ring (33). Release press and remove spring retainer (32) and spring (31). Remove snap ring (30), backing plate (29), friction plates (28),

separator plates (27), piston (24) with seal rings (22 and 25) and "O" rings (23 and 26). Remove bearing cone (1), thrust washers (2), thrust bearing (3), second gear (6), roller bearings (4), spacer (5), thrust washers (7) and thrust bearing (8). Use a press and spring compressor tool (CAS-1992) to compress piston return spring (11) and remove snap ring (9). Remove spring retainer (10), spring (11), snap ring (12), backing plate (13), friction plates (14), separator plates (15), piston (18) with seal rings (16 and 20) and "O" rings (17 and 19) from drop shaft (21).

Clean and inspect all parts and renew any showing excessive wear or other damage. Thickness of new friction plates (14 and 28) is 2.45-2.60 mm (0.096-0.102 in.). Thickness of new separator plates (15 and 27) is 2.11-2.31 mm (0.083-0.091 in.).

Use Figs. 103 and 104 as guides and reassemble drop shaft assembly in reverse order of disassembly, keeping the following points in mind: Use all new "O" ring and seal rings during reassembly. Soak friction plates (14 and 28) in clean Hy-Tran Plus oil before installation. Heat bearing cones (1 and 40) in a bearing oven to a temperature of 150° C (302° F) when installing. Install seal rings (41) on drop shaft (21) using seal expander (CAS-2002), then compress seal rings into grooves in drop shaft using seal compressor (CAS-2005-4).

To reinstall the input shaft and drop shaft assemblies, first make sure rear bearing cups are in place. Install drop shaft assembly in place, then install input shaft assembly. Install front bearing cups in front bearing carrier plate, then install the bearing carrier plate. Install front carrier plate retaining bolts and tighten to a torque of 70-79 N•m (52-58 ft.-lbs.).

Install drop shaft bearing cover (without shims) and install retaining bolts finger tight. Install input shaft bearing cover (without shims) and tighten re-

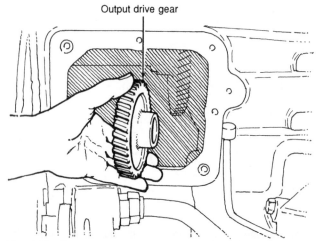

Fig. 100—Remove access cover, then unbolt and remove output drive gear from rear of Powershift drop shaft.

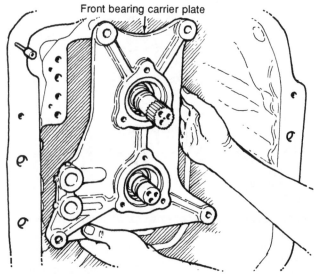

Fig. 101—Remove the four bolts, then carefully remove front bearing carrier plate.

taining bolts to 9 N·m (80 in.-lbs.) while rotating input shaft. Loosen input shaft bearing cover bolts, then retighten the bolts to a torque of 2.8 N·m (25 in.-lbs.) while rotating the shaft. Use a feeler gauge and measure the gap between bearing cover and front bearing carrier plate next to the three mounting bolts. Take an average of the three measurements and add 0.36 mm (0.014 in.). This will be the required shim pack (45—Fig. 102) thickness to be installed. Tag the shim pack for later installation. Shims for the bearing covers are available in thicknesses of 0.076, 0.127. 0.254 and 0.635 mm (0.003, 0.005, 0.010 and 0.025 in.).

Loosen the input shaft bearing cover bolts. Tighten drop shaft bearing cover bolts to 9 N·m (80 in.-lbs.) while rotating the drop shaft. Loosen the retaining bolts, then retighten the bolts to a torque of 2.8 N·m (25 in.-lbs.) while rotating the drop shaft. Use a feeler gauge and measure the gap between the bearing cover and the front bearing carrier plate next to three mounting bolts. Take an average of the three measurements and add 0.25 mm (0.010 in.). This will be the required shim pack (43—Fig. 103) thickness to be installed.

Unbolt and remove input shaft and drop shaft bearing covers. Reinstall covers with the selected

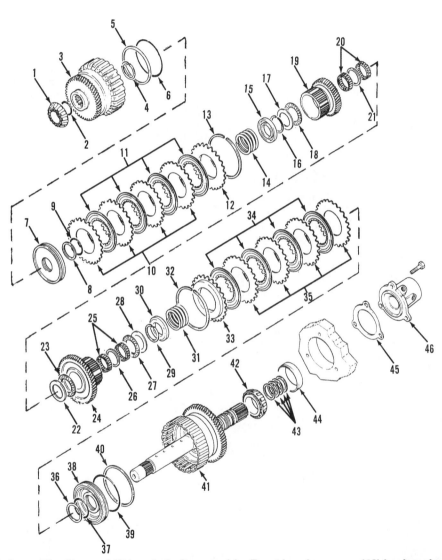

Fig. 102—Exploded view of the Powershift input shaft assembly. Front bearing cover (46) is also shown at (1—Fig. 99).

1. Bearing cone	10. Separator		33. Backing plate	40. Seal ring
2. "O" ring	plates (4)	17. Thrust washer	34. Friction	41. Input shaft &
3. Second drive gear	11. Friction	18. Thrust bearing	plates (4)	fourth drive gear
4. Snap ring	plates (4)	19. First drive gear	35. Separator	42. Bearing cone
5. Seal ring	12. Backing plate	20. Roller bearings	plates (4)	43. Seal rings
6. "O" ring	13. Snap ring	21. Spacer	36. Seal ring	44. Bearing cup
7. Piston	14. Spring	22. Thrust washer	37. "O" ring	45. Shims
8. Seal ring	15. Spring retainer	23. Thrust bearing	38. Piston	46. Front bearing
9. "O" ring	16. Snap ring	24. Third speed gear	39. "O" ring	cover
		25. Roller bearings		
		26. Spacer		
		27. Thrust bearing		
		28. Thrust washer		
		29. Snap ring		
		30. Spring retainer		
		31. Spring		
		32. Snap ring		

shim packs. Tighten bearing cover retaining bolts to a torque of 40-46 N·m (30-34 ft.-lbs.).

Working through the right side access hole, install the drive gear on rear end of drop shaft. Refer to Fig. 105 and measure distance "A" from face of drive gear (1) hub to end of drop shaft (2). Subtract 0.08 mm (0.003 in.) from this measurement to determine the thickness of shim pack to be installed. Shims are available in thicknesses of 0.076, 0.305 and 1.017 mm (0.003, 0.012 and 0.040 in.). Install drive gear retaining bolt and washer with the determined shim pack.

Tighten the bolt to a torque of 101-113 N·m (75-83 ft.-lbs.).

Clean access cover and transmission housing. Apply a bead of Loctite 515 to the face of access cover and install the cover. Tighten retaining bolts securely.

Install clutch supply tubes and lubrication tubes to the input shaft and drop shaft front bearing covers.

Reconnect tractor by reversing the splitting procedure. Fill transmission to correct level with Hy-Tran Plus oil.

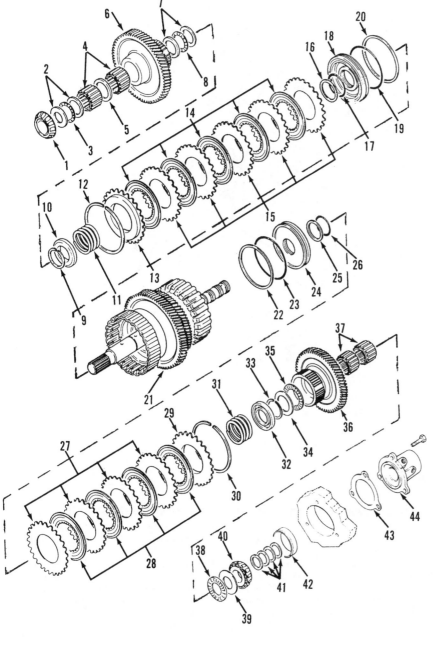

Fig. 103—Exploded view of Powershift drop shaft assembly. Front bearing cover (44) is also shown at (2—Fig. 99).

1. Bearing cone
2. Thrust washers
3. Thrust bearing
4. Roller bearings
5. Spacer
6. Second driven gear
7. Thrust washers
8. Thrust bearing
9. Snap ring
10. Spring retainer
11. Spring
12. Snap ring
13. Backing plate
14. Friction plates (5)
15. Separator plates (5)
16. Seal ring
17. "O" ring
18. Piston
19. "O" ring
20. Seal ring
21. Drop shaft & first & third driven gears
22. Seal ring
23. "O" ring
24. Piston
25. Seal ring
26. "O" ring
27. Separator plates (4)
28. Friction plates (4)
29. Backing plate
30. Snap ring
31. Spring
32. Spring retainer
33. Snap ring
34. Thrust washer
35. Thrust bearing
36. Fourth driven gear
37. Roller bearings
38. Thrust bearing
39. Thrust washer
40. Bearing cone
41. Seal rings
42. Bearing cup
43. Shims
44. Front bearing cover

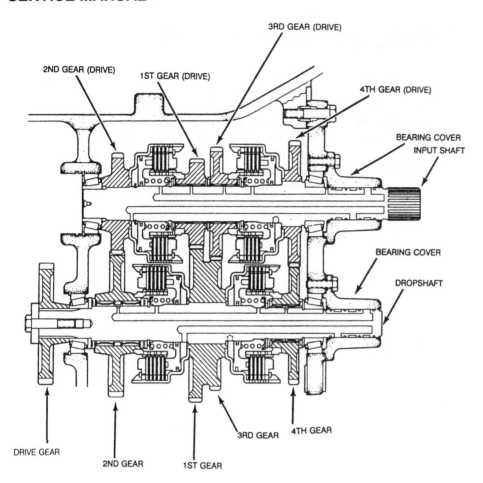

3RD GEAR (DRIVE)

2ND GEAR (DRIVE)

1ST GEAR (DRIVE)

4TH GEAR (DRIVE)

BEARING COVER
INPUT SHAFT

BEARING COVER

DROPSHAFT

DRIVE GEAR

2ND GEAR

1ST GEAR

3RD GEAR

4TH GEAR

Fig. 104—Cross-sectional view of Powershift transmission.

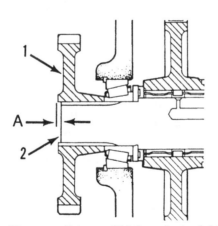

Fig. 105—Measure distance "A" from face of drive gear (1) hub to end of drop shaft (2), then subtract 0.003 inch (0.08 mm) to determine required shim pack to be installed. Refer to text.

SYNCHROMESH TRANSMISSION

All models are available with a 4-speed synchromesh transmission. This transmission, in conjunction with the 4-speed range transmission and the forward-reverse shuttle transmission, provides 16 forward and 12 reverse speeds. Fourth range is not used in reverse. When optional creeper drive is used, 8 low speeds forward and 8 speeds reverse are provided as range 3 and 4 are locked out.

OPERATION

All Models So Equipped

109. Operation of tractors equipped with synchromesh transmission is controlled by three levers as shown in Fig. 106: forward-reverse, synchromesh speed and range. When selecting speeds with tractor stationary, select forward or reverse, fully depress clutch (inching) pedal and select desired range, move speed lever to first speed position, then slowly release clutch pedal. Tractor will start to move.

When selecting speeds with tractor moving, fully depress clutch pedal and move range or speed lever to desired new position. To change direction of travel, always slow tractor speed before moving forward-reverse lever to the desired direction.

For smooth speed changes, move the speed lever through the speeds one at a time in the correct sequence. Neutral is provided in both speed and range transmissions.

R&R AND OVERHAUL

110. To remove the synchromesh transmission, first split tractor between engine and speed transmission as outlined in paragraph 103. Refer to Fig. 107, remove retaining clips and disconnect 1st/2nd (1) and 3rd/4th (2) gear selector rods. Unbolt and remove access cover (Fig. 108) from right side of speed transmission housing. Unbolt and remove the output drive gear from rear of speed transmission drop shaft as shown in Fig. 109. Disconnect the lubrication line

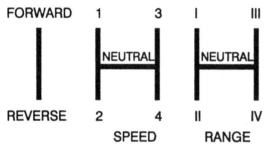

Fig. 106—Control lever shift patterns on models equipped with synchromesh speed transmission.

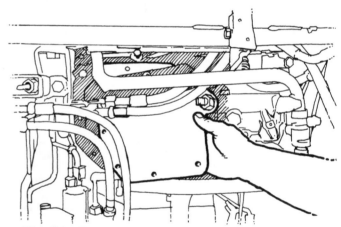

Fig. 108—Remove access cover from right side of speed transmission.

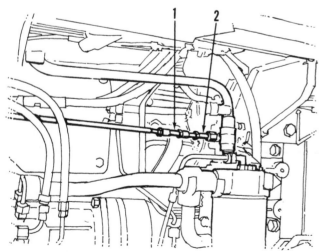

Fig. 107—Disconnect 1st/2nd (1) and 3rd/4th (2) gear selector rods as shown.

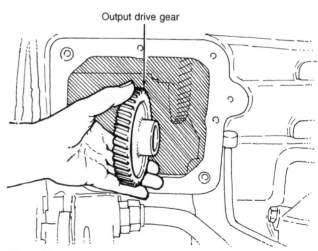

Fig. 109—Unbolt and remove drive gear from rear of synchromesh speed transmission drop shaft.

from drop shaft front bearing cover, then unbolt and remove input shaft and drop shaft front bearing covers and shims. Refer to Fig. 110 and loosen locknut (1) and set screw (2) until the 3rd/4th gear selector fork can be moved on the shift rail, then repeat the operation on the 1st/2nd shift fork. Remove the retaining bolts from bearing carrier plate and install 12 mm dowel studs (1—Fig. 111). Move one shift rail at a time rearward and install detent ball retainers (CAS-1994) as shown in Fig. 111. Carefully remove front bearing carrier plate. Lift out input (upper) shaft assembly and place it on a clean bench. Withdraw the 3rd/4th shift rail (1—Fig. 112) and remove the shift fork (3), then remove 1st/2nd shift rail (2) and fork in the same manner. Lift out drop (lower) shaft assembly (4) and place on bench.

To disassemble the input shaft, refer to Fig. 113 and remove bearing cone (9). Remove snap ring (8) and bearing cone (1). Remove 2nd drive gear (2), spacer (3), 1st drive gear (4), 3rd drive gear (5), spacer (6) and 4th drive gear (7) from input shaft (10).

Clean and inspect all parts and renew any showing excessive wear or other damage. Use Figs. 113 and 114 as guides and reassemble by reversing the disassembly procedure. Heat bearing cones (1 and 9) to a temperature of 150° C (302° F) before installing on input shaft. Lay assembled unit aside in a clean area for later installation.

To disassemble the drop shaft, refer to Figs. 115 and 116, then remove front bearing cone (38). Remove snap ring (36—Fig. 115), thrust washer (35), thrust

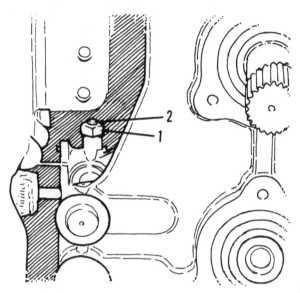

Fig. 110—Loosen locknut (1) and set screw (2) until the 3rd/4th selector fork can be moved on the shift rail. Repeat operation for the 1st/2nd shift fork.

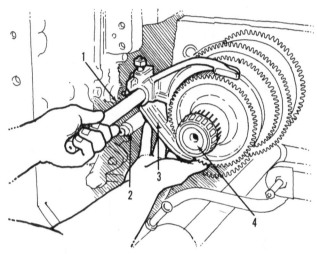

Fig. 112—Support gear selector fork when removing shift fork rail.

1. 3rd/4th shift fork rail
2. 1st/2nd shift fork rail
3. 3rd/4th gear selector fork
4. Drop shaft assy.

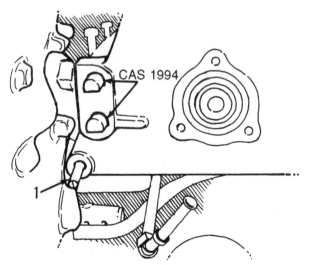

Fig. 111—Unbolt front bearing carrier plate and install 12 mm dowel studs (1) in lower two holes.

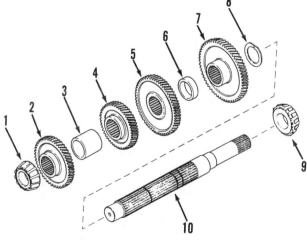

Fig. 113—Exploded view of input shaft assembly.

1. Bearing cone (rear)
2. Drive gear (2nd)
3. Spacer
4. Drive gear (1st)
5. Drive gear (3rd)
6. Spacer
7. Drive gear (4th)
8. Snap ring
9. Bearing cone (front)
10. Input shaft

bearing (34), 4th drive gear (33), thrust bearing (32) and shim (30). Remove 3rd and 4th synchronizer (28) with cups (27 and 29) and hub (31). Remove shim (26), thrust bearing (25), 3rd driven gear (24), thrust bearing (23) and thrust washer (22). Remove snap ring (21), thrust washer (20), thrust bearing (19), 1st driven gear (18), thrust bearing (17) and shim (15). Remove 1st and 2nd synchronizer (13) with cups (12 and 14) and hub (16). Remove shim (11), thrust bear-

ing (10) and thrust washer (9). Remove snap ring (8), 2nd driven gear (7), roller bearings (5), spacers (6), thrust washer (4), thrust bearing (3), rear bearing cone (1) and thrust washer (2). Clean and inspect all parts and renew any showing excessive wear or other damage.

During assembly, synchronizer clearance must be shim adjusted. Shims (11, 15, 26 and 30) are available in thicknesses of 0.50 and 0.75 mm (0.020 and 0.030 in.). Install items (1 through 21) on drop shaft (37), using one 0.50 mm (0.020 in.) shim at (11) and (15). Stand shaft in vertical position (rear end down). Refer to Fig. 117 and hold 1st driven gear upward until the thrust washer (20—Fig. 115) is against snap ring (21). Using a feeler gauge, take three evenly spaced measurements between synchronizer cup (14—Fig. 117) and shim (15). Average the three measurements for dimension (X). Disassemble and reassemble using correct shim packs divided between (A and B) as described in chart (Fig. 117).

Assemble items (22 through 36—Fig. 115) over front end of drop shaft using one 0.50 mm (0.020 in.)

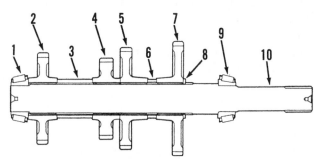

Fig. 114—Cross-sectional view of input shaft assembly. Refer to Fig. 113 for legend.

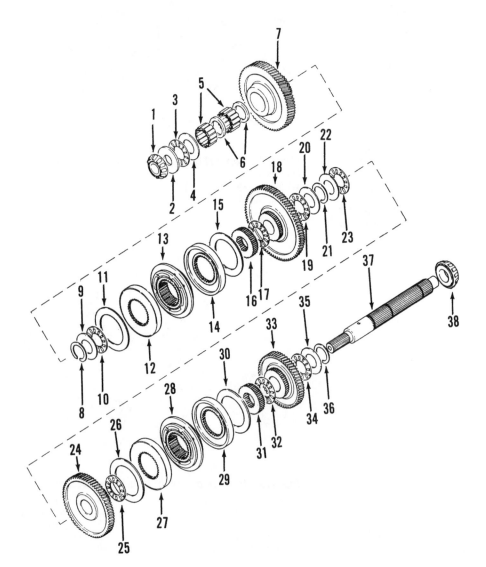

Fig. 115—Exploded view of the drop shaft assembly.

1. Bearing cone (rear)
2. Thrust washer
3. Thrust bearing
4. Thrust washer
5. Roller bearings
6. Spacers
7. Driven gear (2nd)
8. Snap ring
9. Thrust washer
10. Thrust bearing
11. Shim
12. Cup
13. Synchronizer (1st & 2nd)
14. Cup
15. Shim
16. Hub
17. Thrust bearing
18. Driven gear (1st)
19. Thrust bearing
20. Thrust washer
21. Snap ring
22. Thrust washer
23. Thrust bearing
24. Driven gear (3rd)
25. Thrust bearing
26. Shim
27. Cup
28. Synchronizer (3rd & 4th)
29. Cup
30. Shim
31. Hub
32. Thrust bearing
33. Driven gear (4th)
34. Thrust bearing
35. Thrust washer
36. Snap ring
37. Drop shaft
38. Bearing cone (front)

thick shim at (26) and (30). Refer to Fig. 118 and hold 4th driven gear (33) upward until the thrust washer (35—Fig. 115) is against snap ring (36). Using a feeler gauge, take three evenly spaced measurements at (C—Fig. 118) between synchronizer cup (29) and shim (30). Average the three measurements for dimension (Y). Use the chart in Fig. 118 to determine shim packs to be installed at (C and D). Disassemble and reassemble using correct shim packs. Install front bearing cone (38—Fig. 116).

Before installing input and drop shaft assemblies, inspect and renew front and rear bearing cups as required.

Install the drop shaft assembly. Install the 1st/2nd gear selector fork and shift rail, but do not tighten set screw at this time. Repeat the operation for the 3rd/4th gear selector fork and shift rail. Install the input shaft assembly, then install the front bearing carrier plate. Pull the 1st/2nd (lower) shift rail into lower hole of carrier plate until the interlock pin drops into the groove of the shift rail. Remove the ball detent retainer CAS-1994. Turn the second ball detent retainer 180° counterclockwise. Pull 3rd/4th shift rail into the same position as the 1st/2nd shift rail, then remove the ball detent retainer. Remove the two 12 mm dowel studs and install and tighten front bearing carrier plate retaining bolts to a torque of 70-79 N•m (52-58 ft.-lbs.). Tighten the set screws (2—Fig. 110) and locknuts (1) to position selector forks on the shift rails.

Install the drop shaft front bearing cover without shims. Install, but do not tighten, cover retaining bolts. Install input shaft front bearing cover without shims. Install and tighten retaining bolts to a torque of 9 N•m (80 in.-lbs.) while rotating input shaft. Loosen, then retighten the bolts (while rotating the shaft) to a torque of 2.8 N•m (25 in.-lbs.). Use a feeler gauge and measure the gap between the bearing cover and the front bearing carrier plate next to the three mounting bolts. Take an average of the three measurements and add 0.36 mm (0.014 in.) to determine the required shim pack to be installed. Record this thickness and loosen input shaft bearing cover bolts.

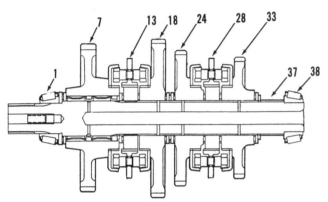

Fig. 116—Cross-sectional view of drop shaft assembly used in synchronized transmission. Refer to Fig. 115 for legend.

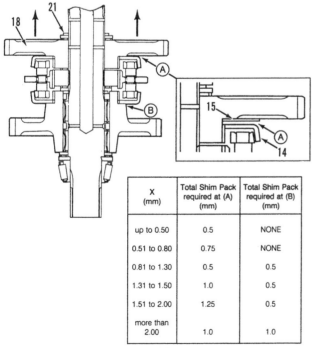

X (mm)	Total Shim Pack required at (A) (mm)	Total Shim Pack required at (B) (mm)
up to 0.50	0.5	NONE
0.51 to 0.80	0.75	NONE
0.81 to 1.30	0.5	0.5
1.31 to 1.50	1.0	0.5
1.51 to 2.00	1.25	0.5
more than 2.00	1.0	1.0

Fig. 117—Measure gap between synchronizer cup (14) and shim (15), then use the chart to determine correct total shim packs (A and B) to be installed. Refer to text.

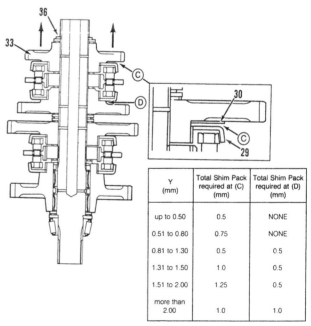

Y (mm)	Total Shim Pack required at (C) (mm)	Total Shim Pack required at (D) (mm)
up to 0.50	0.5	NONE
0.51 to 0.80	0.75	NONE
0.81 to 1.30	0.5	0.5
1.31 to 1.50	1.0	0.5
1.51 to 2.00	1.25	0.5
more than 2.00	1.0	1.0

Fig. 118—Measure gap between synchronizer cup (29) and shim (30), then use the chart to determine correct total shim packs (C and D) to be installed. Refer to text.

Shims for the bearing covers are available in thicknesses of 0.076, 0.127, 0.254 and 0.635 mm (0.003, 0.005, 0.010 and 0.025 in.).

Tighten drop shaft bearing cover bolts to 9 N·m (80 in.-lbs.) while rotating the drop shaft. Loosen the retaining bolts, then retighten to a torque of 2.8 N·m (25 in.-lbs.) while rotating the drop shaft. Use a feeler gauge and measure the gap between the bearing cover and the front bearing carrier plate next to the three bolts. Take an average of the three measurements and add 0.36 mm (0.014 in.). This will be the required shim pack thickness to be installed.

Unbolt and remove input shaft and drop shaft bearing covers. Reinstall covers with the selected shim packs. Tighten bearing cover retaining bolts to a torque of 40-46 N·m (30-34 ft.-lbs.).

Working through the right side access hole, install the drive gear on rear end of drop shaft. Refer to Fig. 119 and measure distance "A" from face of drive gear (1) hub to end of drop shaft (2). Subtract 0.08 mm (0.003 in.) from this measurement to determine the thickness of shim pack to be installed. Shims are available in thicknesses of 0.076, 0.305 and 1.016 mm (0.003, 0.012 and 0.040 in.). Install drive gear retaining bolt and washer with the determined shim pack. Tighten the bolt to a torque of 101-113 N·m (75-83 ft.-lbs.).

Clean access cover and transmission housing. Apply a bead of Loctite 515 to face of access cover and install the cover. Tighten retaining bolts securely.

Connect the lubrication line to the drop shaft front bearing cover. Reconnect tractor by reversing the splitting procedure. Fill transmission to correct level with Hy-Tran Plus oil.

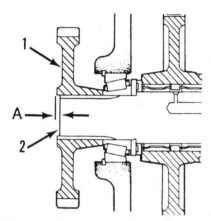

Fig. 119—Measure distance "A" from face of drive gear hub (1) to end of drop shaft (2), then subtract 0.003 inch (0.08 mm) to determine required shim pack to be installed. Refer to text.

RANGE TRANSMISSION AND FORWARD-REVERSE SHUTTLE TRANSMISSION

All models are equipped with a synchronized, 4-speed, range transmission and a hydraulic, clutch-operated, forward-reverse shuttle transmission. The forward-reverse transmission is located directly above the range transmission. Both transmissions are located in the front portion of the rear main frame. Shift control lever for the range transmission is located on the right side console and the forward-reverse control lever is located on the instrument panel.

OPERATION

All Models

111. When selecting a range speed with tractor stationary, fully depress clutch (inching) pedal, then move shift lever to desired range speed. When shifting "on-the-go," depress clutch pedal and shift control lever to the next (higher or lower) range speed.

To change direction of travel, always slow tractor speed before moving forward-reverse lever to the desired direction. Clutch (inching) pedal does not need to be depressed when shifting to forward or reverse.

R&R AND OVERHAUL

All Models

112. To remove the range transmission and the forward-reverse drive, first separate tractor between the speed transmission and the range transmission

as follows: Block front wheels securely and place wood wedges between front axle and front support to prevent tipping. Disconnect battery cables. Remove plugs and drain transmission oil. Unbolt and remove seat assembly. Place identification tags on electrical connectors, then disconnect the control valve solenoids, pressure switches and ground leads (Fig. 138). Disconnect and cap hydraulic tubes as required. Disconnect the clutch cable and the park brake cable. Remove the three bolts from bottom of transmission and install the transmission splitting rail (CAS-10853) under range transmission. Place suitable blocks under speed transmission. Remove rear wheels. Unbolt and remove left step assembly. Drain the fuel tank and disconnect fuel supply hose, return hose and breather hose. Unbolt and remove fuel tank. Install three cab stands (CAS-3389) under cab. Disconnect pto control cable, then remove cable and switch from bracket. Disconnect ground speed sensor electrical connector. Disconnect remote valve control rods and remove remote valve return tube. Disconnect and cap brake hoses. Unbolt and remove protection shield from right side of transmission. Disconnect and cap the hydraulic lines from front-wheel drive valve, if so equipped, and the compensator valve. Raise hood, remove exhaust extension tube and remove rear hood panel. Disconnect ground cables from both sides of cab. Remove rear cab mounting nuts and loosen front cab mounting nuts. Raise rear of cab to clear cab mounts. Check to be sure all hydraulic lines and wires are disconnected for the split. Remove the speed to range transmission mounting bolts and carefully separate the tractor.

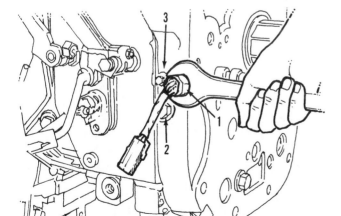

Fig. 120—Removal of the fourth range switch (1), 1st/2nd detent plug (2) and 3rd/4th detent plug (3). Refer to text.

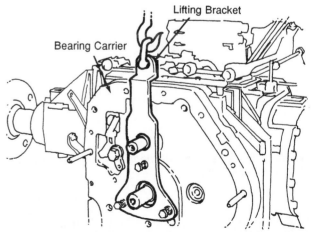

Fig. 121—Using a hoist and lifting bracket (CAS-1993), remove bearing carrier.

On two-wheel-drive models, remove park brake assembly, pinion bolt, washer, shims and spacer. On four-wheel-drive models, remove pinion boot, washer, shims, clutch drive gear, idler gear and the front-drive clutch and park brake assembly. On all models, remove the park brake gear hub.

Install alignment studs (CAS-1995) at each side of bearing carrier housing. Refer to Fig. 120, loosen locknut and remove fourth range switch (1). Remove 1st/2nd detent plug (2) and 3rd/4th detent plug (3), then remove spacers, springs and detent balls. Remove the three bearing carrier mounting bolts. Install a lifting bracket (CAS-1993) to the carrier housing using the park brake holes as shown in Fig. 121. Connect a hoist to the lifting bracket. Using a

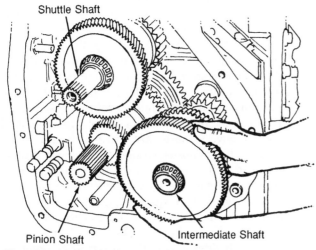

Fig. 122—View showing the intermediate shaft assembly being removed from range transmission.

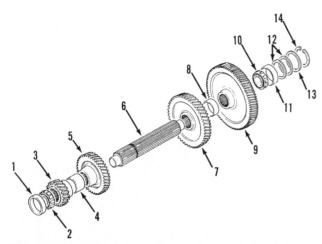

Fig. 123—Exploded view of range transmission intermediate shaft assembly.

1. Bearing cup
2. Bearing cone (rear)
3. Drive gear (1st)
4. Wide spacer
5. Drive gear (2nd)
6. Intermediate shaft
7. Drive gear (3rd)
8. Narrow spacer
9. Drive gear (4th)
10. Bearing cone (front)
11. Bearing cup
12. Shims
13. Spacer
14. Snap ring

suitable puller, pull the bearing carrier off the pinion shaft. Remove the bearing carrier with the lifting bracket. Remove the guide studs. Lift out the intermediate shaft assembly as shown in Fig. 122. Withdraw the pto shaft from lower front of range transmission.

To disassemble the range intermediate shaft, place assembly in a press (4th gear up). Support under 3rd gear. Remove front bearing cone (10—Fig. 123), 4th drive gear (9), spacer (8) and 3rd drive gear (7). Remove 2nd drive gear (5) and spacer (4). Turn shaft over and press shaft (6) from rear bearing cone (2) and 1st drive gear (3). Clean and inspect all parts for wear or other damage and renew as required. If rear bearing (2) is to be renewed, use a slide hammer puller to remove bearing cup (1) from transmission housing. If front bearing cone (10) is to be renewed, remove snap ring (14), spacer (13), shims (12) and bearing cup (11) from bearing carrier housing.

When reassembling, heat gears and bearings in a bearing oven to a temperature of 150° C (302° F). Using Fig. 123 and Fig. 124 as a guide, install bearings, gears and spacers on the shaft. Place the assembly in a press and apply pressure to end of bearings. Make sure all parts are in contact and that spacers cannot be turned. If removed, install new rear bearing cup (1—Fig. 123) until seated against shoulder in housing. If removed, install front bearing cup (11), spacer (13) and snap ring (14). Do not install shims (12) at this time. Shims can be installed only when bearing carrier is installed and the intermediate shaft end play is determined. Make sure bearing cup, spacer and snap ring are in contact. Lay intermediate shaft aside for later installation.

Place shift forks in neutral position and place identification marks on forks for aid in reassembly. Loosen locknuts and remove Allen screws (Fig. 125). Remove the 3rd/4th (upper) shift rail, support the shift forks and remove 1st/2nd shift rail. Remove the 3rd/4th shift fork and remove interlock pin from the interlock housing. Then, remove 1st/2nd shift fork.

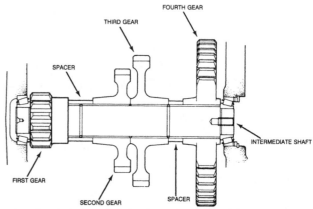

Fig. 124—Cross-sectional view showing correct assembly of the intermediate shaft components.

Mark position of shift levers to outer end of shifter pivot shafts (2 and 3—Fig. 126). Remove clamping bolts and slide off shift levers. Loosen bushing locknuts and unscrew bushings (4). Remove shifter pivot shafts from inside the housing. Remove and discard "O" rings from bushings and pivot shafts.

NOTE: To remove the shuttle (forward-reverse) shaft, some disassembly of the shuttle shaft and pinion shaft components is required within the range housing.

Using a suitable puller, remove front bearing cone (56—Fig. 127) from shuttle shaft. Remove snap ring (55), thrust washer (54), thrust bearing (53) and forward gear (50) with roller bearings (51), spacer (52) and seal ring (49). Remove thrust bearing (48) and thrust washer (47). Remove pinion shaft snap ring (40—Fig. 128), thrust washer (39), thrust bearing (38), 4th driven gear (37) and spacer (36). Remove 3rd/4th synchronizer assembly (32) and shim spacers (31 and 33). Remove thrust bearing (35) and synchronizer hub (34). Remove snap ring (30), thrust washer (29), thrust bearing (28), 3rd driven gear (27), thrust bearing (26) and thrust washer (25). Remove snap ring (24), thrust washer (23) and thrust bearing (22). Withdraw the shuttle shaft assembly. The 2nd driven gear (21) will slide forward on pinion shaft when shuttle shaft is being removed.

To disassemble the shuttle shaft, remove the forward clutch snap ring (41—Fig. 127), then remove backing plate (40), friction discs (39) and separator plates (38). Place shuttle shaft in a press. Using spring compressor sleeve (CAS-1992), compress piston return spring (44) and remove snap ring (46) from its groove in shuttle shaft. Remove shuttle shaft from press and remove snap ring (46), spring retainer (45), spring (44), lubrication sleeve (43) and shim spacer (42). Bump front end of shuttle shaft on a block of wood to remove forward piston (35) from clutch housing. Remove and discard seal rings (34 and 36) and

"O" rings (33 and 37) from piston. Remove the four seal rings (1) from rear of shaft. Use a suitable puller and remove rear bearing cone (3). Remove snap ring (4), thrust washer (5), thrust bearing (6) and reverse gear (9) with seal ring (10), roller bearings (7) and spacer (8). Remove thrust bearing (11) and thrust washer (12). Remove reverse clutch snap ring (19), backing plate (20), friction discs (21) and separator plates (22). Place shuttle shaft in a press. Using spring compressor sleeve (CAS-1992), compress piston return spring (15) and remove snap ring (13) from its groove in shuttle shaft. Remove shuttle shaft from press and remove snap ring (13), spring retainer (14), spring (15), lubrication sleeve (16) and shim spacer (18). Bump rear end of shuttle shaft on a block of wood to remove reverse piston (25) from clutch housing. Remove and discard seal rings (24 and 26) and "O" rings (23 and 27) from piston.

Clean and inspect all parts and renew any showing excessive wear or other damage. Check thickness of friction discs and separator plates. Renew friction discs if thickness is less than 2.4 mm (0.094 in.). Renew separator plates if thickness is less than 2.1 mm (0.082 in.). Check shuttle shaft wear sleeve (Fig. 129) for excessive wear. Renew sleeve if bore size is more than 36.07 mm (1.42 in.). Oil holes in sleeve and housing must be aligned. Make certain oil passages in shuttle shaft are clean and that passage plugs in ends of shaft are correctly installed.

During reassembly of the shuttle shaft, use all new "O" rings and seal rings and lubricate all parts with clean Hy-Tran Plus oil. Install piston (35—Fig. 127), with grooved side facing outward, in forward clutch housing until seated. Install lubrication sleeve (43) and hold sleeve firmly against the tapered seat on shaft. Refer to Fig. 130 and measure gap between lubrication sleeve flange and forward piston. Use chart in Fig. 130 to determine correct thickness of shim spacer (42—Fig. 127) to be installed. Shim spac-

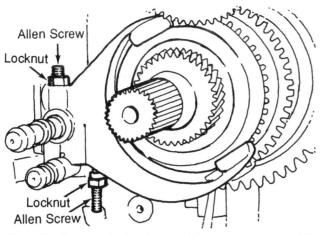

Fig. 125—Remove locknuts and Allen screws from shift forks.

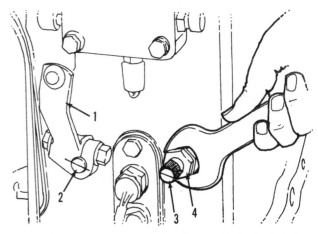

Fig. 126—Mark position of shift levers (3rd/4th shift lever is removed) on pivot shafts (2 and 3) to aid reassembly.

1. 1st/2nd shift lever	3. 3rd/4th pivot shaft
2. 1st/2nd pivot shaft	4. Bushing

ers are available in thicknesses of 0.75, 1.00, 1.25 and 2.25 mm (0.029, 0.039, 0.049 and 0.089 in.). Remove the sleeve and install the determined shim spacer, then reinstall the sleeve. The gap at this time must be between 0.07-0.83 mm (0.002 and 0.032 in.). Install piston return spring (44), spring retainer (45) and snap ring (46) onto the shuttle shaft. Use a press and spring compressor sleeve (CAS-1992) to compress the spring until snap ring can be installed into the groove in shaft. Starting with a separator plate (39) and alternating with a friction disc (38), install six separator plates and six friction discs. Install backing plate (40) and snap ring (41). Items (47 through 56) will be assembled during shuttle shaft installation.

Assemble items (13 through 27) in reverse clutch housing in the same manner. Then install thrust washer (12) and thrust bearing (11) with open (needle bearing) side toward the thrust washer. Install new seal ring (10) on reverse gear (9). Install roller bearings (7) with spacer (8) between the bearings, onto the shaft. Align friction discs and install reverse gear, making sure that all discs engage the gear hub splines. Install thrust bearing (6) and thrust washer (5) so that open side of thrust bearing is toward thrust washer. Install snap ring (4). Heat bearing cone (3) in a bearing oven to a temperature of 150° C (302° F) and install on shaft until seated against the snap ring. Use seal expander (CAS-1999) and install the four seal rings (1). Use seal compressor (CAS-2005-1) to compress seal rings in their grooves to correct size.

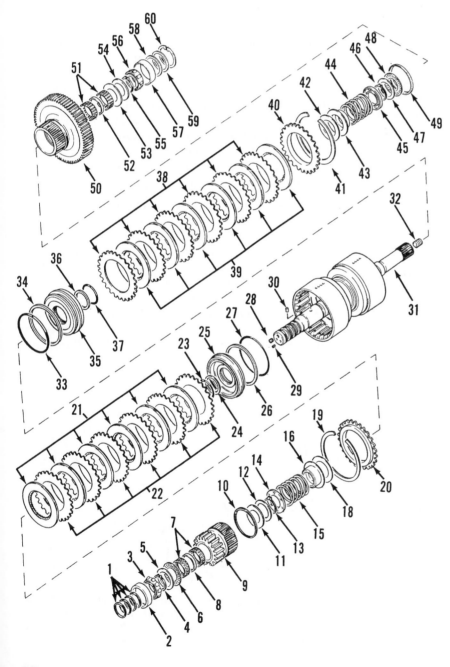

Fig. 127—Exploded view of the forward-reverse shuttle shaft assembly used on all models.

1. Seal rings (4)	32. Plug
2. Bearing cup	33. "O" ring
3. Bearing cone (rear)	34. Seal ring
4. Snap ring	35. Piston (forward)
5. Thrust washer	36. Seal ring
6. Thrust bearing	37. "O" ring
7. Roller bearing	38. Separator plates (6)
8. Spacer	39. Friction discs (6)
9. Reverse gear	40. Backing plate
10. Seal ring	41. Snap ring
11. Thrust bearing	42. Shim spacer
12. Thrust washer	43. Lubrication sleeve
13. Snap ring	44. Spring
14. Spring retainer	45. Spring retainer
15. Spring	46. Snap ring
16. Lubrication sleeve	47. Thrust washer
18. Shim spacer	48. Thrust bearing
19. Snap ring	49. Seal ring
20. Backing plate	50. Forward gear
21. Friction discs (6)	51. Roller bearings
22. Separator plates (6)	52. Spacer
23. "O" ring	53. Thrust bearing
24. Seal ring	54. Thrust washer
25. Piston (reverse)	55. Snap ring
26. Seal ring	56. Bearing cone (front)
27. "O" ring	57. Bearing cup
28. Plug	58. Shims
29. Plug (2)	59. Spacer
30. Roll pin	60. Snap ring
31. Shuttle shaft	

If a new bearing cone (3) was installed, install new bearing cup (2) into housing until seated against shoulder. Lay the assembly aside for later installation. If new bearing cone (56) is to be installed, remove snap ring (60), spacer (59), shims (58) and bearing cup (57) from bearing carrier housing. Install new bearing cup, spacer and snap ring in the bearing carrier. Do not install shims (58) at this time. Shims can be installed only when bearing carrier is installed and the shuttle shaft end play is determined.

To remove the pinion shaft (1—Fig. 128), first remove the differential assembly as outlined in paragraph 122. Refer to Fig. 131 and remove the pinion shaft lubrication tubes. Using blocks of wood, hold

pinion shaft forward. Then, with the forward-reverse shuttle shaft removed, remove 2nd driven gear (21—Fig. 128) from pinion shaft. Remove 1st/2nd synchronizer (18) with shim spacers (17 and 19). Remove thrust bearing (20) and synchronizer hub (16). Remove snap ring (15), thrust washers (13) and thrust bearing (14). Remove 1st driven gear (12), spacer (11) and roller bearings (9) with spacer (10). Remove thrust washers (6 and 8) and thrust bearing (7). Remove wood blocks, then remove pinion shaft (1) and bearing cone (3) from rear of housing. Press shaft from bearing cone. Drive bearing cup (4) with shim (5) from housing.

Clean and inspect all parts and renew any showing excessive wear or other damage. Drive pinion shaft (1) and ring gear (2) are available only as a matched set. Refer to paragraph 122 for ring gear renewal procedure. To calculate the correct thickness of pinion shaft mounting distance shim pack (5—Fig. 128), subtract the dimension etched in millimeters on rim of ring gear from the dimension stamped in upper left face in rear of main frame. Then subtract the bearing growth constant of 0.04 mm. The result will be the required shim pack thickness. Shims (5) are available in thicknesses of 0.08, 0.18, 0.25 and 0.51 mm (0.003, 0.007, 0.010 and 0.020 in.). Install the determined shim pack (5) and pinion bearing cup (4) into the transmission housing. Heat bearing cone (3) to a temperature of 150° C (302° F) and install on pinion shaft until seated against shoulder. Install pinion shaft into transmission housing and hold pinion in forward position with blocks of wood. If removed, install bearing cup (41) in the bearing carrier housing. Install two 16 mm alignment studs into the transmission housing. Using a hoist and lifting bracket (CAS-1993), install bearing carrier housing. Install retaining bolts with hardened washers and

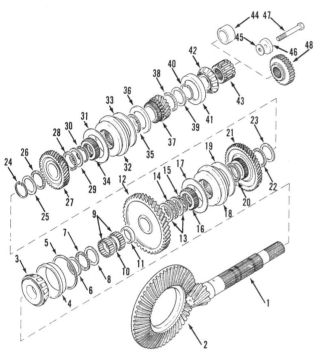

Fig. 128—Exploded view of pinion shaft assembly used on all models. Pinion shaft (1) and ring gear (2) are available only as a matched set.

1. Pinion shaft	25. Thrust washer
2. Ring gear	26. Thrust bearing
3. Bearing cone	27. Driven gear (3rd)
4. Bearing cup	28. Thrust bearing
5. Shim	29. Thrust washer
6. Thrust washer	30. Snap ring
7. Thrust bearing	31. Shim spacer
8. Thrust washer	32. Synchronizer
9. Roller bearings	33. Shim spacer
10. Spacer	34. Hub
11. Spacer	35. Thrust bearing
12. Driven gear (1st)	36. Spacer
13. Thrust washers	37. Drive gear (4th)
14. Thrust bearing	38. Thrust bearing
15. Snap ring	39. Thrust washer
16. Hub	40. Snap ring
17. Shim spacer	41. Bearing cup
18. Synchronizer	42. Bearing cone
19. Shim spacer	43. Park brake hub
20. Thrust bearing	44. Spacer (two-wheel drive)
21. Driven gear (2nd)	45. Shim
22. Thrust bearing	46. Washer
23. Thrust washer	47. Bolt
24. Snap ring	48. Drive gear (four-wheel drive)

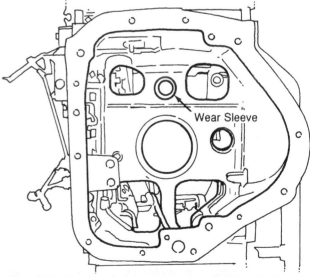

Fig. 129—Renew forward-reverse shuttle shaft wear sleeve if bore size is more than 1.42 inch (36.07 mm).

tighten to a torque of 235-265 N·m (175-195 ft.-lbs.). Remove the lifting bracket.

Install a master bearing cone (CAS-2024) onto the pinion shaft. Install park brake hub (43) and spacer (44) or four-wheel drive gear (48). Refer to Fig. 132 and install set up washer (CAS-1998) and tighten bolt to a torque of 20 N·m (180 in.-lbs.) Remove wood blocks from rear of pinion shaft. Rotate pinion shaft one revolution and recheck mounting bolt torque. Using a depth gauge, measure the distance from face of set up washer to end of pinion shaft through both holes in the washer. Subtract the set up washer thickness from the average of the two measurements. Then, add 0.043 mm (0.0017 in.). This will give the correct shim pack (45—Fig. 128) thickness to be installed. Make a note of the shims required. Shims are available in thicknesses of 0.077, 0.305 and 1.017 mm (0.003, 0.012 and 0.040 in.).

Block pinion shaft forward and remove set up washer, spacer (44—Fig. 128) or gear (48), park brake hub (43) and master bearing cone (CAS-2024). Attach

lifting bracket and remove bearing carrier housing. Install the thick thrust washers (6 and 8) with thrust bearing (7) between the washers. Install roller bearings (9) with spacer (10) between bearings. Install spacer (11), 1st driven gear (12), thin thrust washers (13) with thrust bearing (14) between washers, then install snap ring (15). Install synchronizer hub (16) and thrust bearing (20). Make sure the recessed side of hub is toward the snap ring.

Remove blocks from rear of pinion shaft and install the pinion shaft lubrication tubes (Fig. 131). Refer to paragraph 122 and reinstall the differential assembly. Install synchronizer (18—Fig. 128) and shims (17 and 19). Use one 0.5 mm (0.020 in.) thick shim at (17 and 19). Install 2nd driven gear (21), thrust bearing (22), thrust washer (23) and snap ring (24). Push 2nd driven gear (21) rearward to eliminate the space between the 1st driven gear (12) and the synchronizer assembly, then pull 2nd gear forward. Using a feeler gauge, take three evenly spaced measurements between the synchronizer (18—Fig. 128) and shim (19).

MEASURED GAP		REQUIRED SPACER(S) THICKNESS	
MM	INCH	MM	INCH
4.45 to 4.67	0.17 to 0.183	4.25	0.167
4.19 to 4.42	0.165 to 0.179	4.00	0.157
3.94 to 4.17	0.155 to 0.164	3.75	0.147
3.68 to 3.91	0.145 to 0.154	3.50	0.138
3.43 to 3.65	0.135 to 0.144	3.25	0.128
3.18 to 3.40	0.125 to 0.134	3.00	0.118
2.92 to 3.15	0.115 to 0.124	2.75	0.108
2.67 to 2.90	0.105 to 0.114	2.50	0.098
2.41 to 2.64	0.095 to 0.104	2.25	0.089
2.16 to 2.39	0.085 to 0.094	2.00	0.079
1.91 to 2.13	0.075 to 0.084	1.75	0.069
1.65 to 1.88	0.065 to 0.074	1.50	0.059
1.38 to 1.63	0.054 to 0.064	1.25	0.049
1.21 to 1.37	0.048 to 0.053	1.00	0.039
0.84 to 1.19	0.033 to 0.047	0.75	0.029
0.07 to 0.83	0.002 to 0.032	NO SPACER	

Fig. 130—View showing gap to be measured and the chart used to determine thickness of shim spacer to be installed between lubrication sleeve and clutch piston.

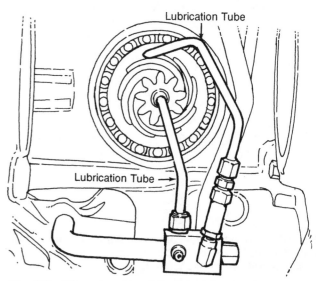

Fig. 131—View showing pinion shaft lubrication tubes.

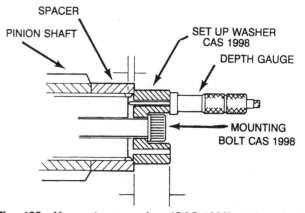

Fig. 132—Use set up washer (CAS-1998) and a depth gauge to determine shim pack thickness to be installed between end of pinion shaft (1—Fig. 128) and washer (46).

Average the three measurements, then use the chart in Fig. 133 to determine the total shim pack required at (17 and 19—Fig. 128). Make a note of the shim packs to be installed.

Install thrust washer (25), thrust washer (26), 3rd driven gear (27), thrust bearing (28), thrust washer (29) and snap ring (30). Install synchronizer hub (34) and thrust bearing (35). Make sure recessed side of hub is facing the snap ring. Install synchronizer (32) and shims (31 and 33). Use one 0.5 mm (0.020 in.) thick shim at (31 and 33). Install spacer (36), 4th driven gear (37), thrust bearing (38), thrust washer

A (mm)	Total Shim Pack required at (17) (mm)	Total Shim Pack required at (19) (mm)
up to 0.50	0.5	NONE
0.51 to 0.80	0.75	NONE
0.81 to 1.30	0.5	0.5
1.31 to 1.50	1.0	0.5
1.51 to 2.00	1.25	0.5
more than 2.00	1.0	1.0

Fig. 133—Chart used to determine required thickness of shim packs (17 and 19—Fig. 128). Dimension "A" is measured between synchronizer (18) and shim (19).

B (mm)	Total Shim Pack required at (31) (mm)	Total Shim Pack required at (33) (mm)
up to 0.50	0.5	NONE
0.51 to 0.80	0.75	NONE
0.81 to 1.30	0.5	0.5
1.31 to 1.50	1.0	0.5
1.51 to 2.00	1.25	0.5
more than 2.00	1.0	1.0

Fig. 134—Chart used to determine required thickness of shim packs (31 and 33—Fig. 128). Dimension "B" is measured between synchronizer (32) and shim (33).

(39) and snap ring (40). Push 4th gear (37) rearward to eliminate the space between 3rd driven gear (27) and the synchronizer assembly, then pull 1st gear forward. Using a feeler gauge, take three evenly spaced measurements between the synchronizer (32—Fig. 128) and shim (33). Average the three measurements, then use the chart in Fig. 134 to determine the total shim pack required at (31 and 33—Fig. 128). Make a note of shim packs to be installed.

NOTE: Shims for (17, 19, 31 and 33) are available in thicknesses of 0.5 and 0.75 mm (0.020 and 0.030 in.).

Remove items (17 through 40) from pinion shaft. Install synchronizer (18) with the previously determined shims (17 and 19). If removed, install thrust bearing (20), then slide 2nd drive gear (21) half way on the pinion shaft. Slide 2nd driven gear rearward as the shuttle shaft is being installed (Fig. 135). Install thrust bearing (22—Fig. 128), thrust washer (23) and snap ring (24). Install thrust washer (25), thrust bearing (26), 3rd driven gear (27), thrust bearing (28), thrust washer (29) and snap ring (30). Install synchronizer hub (34) and thrust bearing (35). Make sure recessed side of hub is facing the snap ring. Install synchronizer (32) with previously determined shims (31 and 33). Install spacer (36), 4th driven gear (37), thrust bearing (38), thrust washer (39) and snap ring (40).

Refer to Fig. 127 and install thrust washer (47), thrust bearing (48) with open (needle) side of bearing toward thrust washer and roller bearings (51) with spacer (52) between the bearings. Install new seal ring (49) on forward gear, align friction discs (39) and install forward gear (50). Make sure all friction discs engage the forward gear hub splines. Install thrust bearing (53) and thrust washer (54), making sure open (needle) side of bearing faces thrust washer, then install snap ring (55). Heat bearing cone (56) in

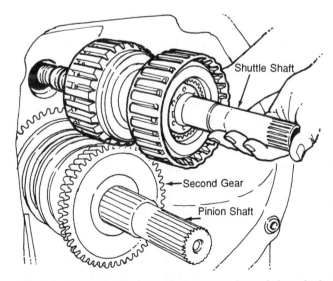

Fig. 135—Move 2nd driven gear rearward on pinion shaft as shuttle shaft is being installed.

a bearing oven to 150° C (302° F) and install bearing cone until seated against shoulder.

Install new "O" rings on shifter pivot shafts and bushings and lubricate with petroleum jelly. Install 1st/2nd pivot bushing about two turns into housing, then install 1st/2nd pivot shaft (from inside) into bushing. Turn bushing clockwise until thread begins to show on inside of housing. Repeat the operation on 3rd/4th pivot shaft and bushing. Install the 1st/2nd shift fork, then the 3rd/4th shift fork. Support shift forks in position and install 1st/2nd shift rail so that Allen screw detent is at bottom. Install the interlock pin into the interlock housing. Install 3rd/4th shift rail so that Allen screw detent is at top. Install Allen screws and tighten locknuts. Turn pivot shaft bushings clockwise until the bushings, pivot shafts and shift forks are in contact. Then, turn bushings counterclockwise until a gap of 0.5-1.0 mm (0.020-0.040 in.) can be measured between pivot shafts and shift forks. Tighten the bushing locknuts. Install external shift levers, align the marks (installed during removal) and tighten clamping bolts. Install new lubrication pipe seal rings and install pto and lube pump drive shaft. Clean mating surfaces of transmission housing and bearing carrier and apply a bead of Loctite 515 to face of transmission. Install the intermediate shaft assembly. Install two 16 mm alignment studs in transmission housing, then using a hoist and lifting bracket, install the bearing carrier. Install bearing carrier mounting bolts with new hardened washers and tighten bolts to a torque of 235-265 N·m (175-195 ft.-lbs.). Remove the lifting bracket.

Heat bearing cone (42—Fig. 128) to a temperature of 150° C (302° F) and install on pinion shaft. Install park brake gear hub (43) with chamfer facing out. Install spacer (44) or front-wheel drive gear (48). Install bolt (47), washer (46) and the previously determined shim pack (45). Tighten bolt to a torque of 335-375 N·m (247-277 ft.-lbs.). Using a dial indicator, measure end play of the shuttle shaft. Subtract 0.06 mm (0.0024 in.) from the end play measurement to determine the required thickness of shim pack (58—Fig. 127). Remove snap ring (60) and spacer (59). Install the required shim pack, spacer and snap ring. Recheck shuttle shaft end play, which should be 0.022-0.102 mm (0.001-0.004 in.).

Install a 10 mm bolt with locknut into front end of intermediate shaft and tighten locknut against end of shaft. Using a dial indicator against bolt head, measure shaft end play. Subtract 0.06 mm (0.0024 in.) from the end play measurement to determine the required thickness of shim pack (12—Fig. 123). Remove snap ring (14) and spacer (13). Install the required shim pack, spacer and snap ring. Recheck intermediate shaft end play, which should be 0.022-0.102 mm (0.001-0.004 in.). Remove bolt and locknut.

Install 3rd/4th (upper) detent ball, spring and spacer, then install plug with new "O" ring. Repeat the operation for the 1st/2nd (lower) detent assembly. Install the 4th range switch.

Reinstall the park brake assembly and on models so equipped, install the front-wheel drive idler gear and clutch.

The balance of reassembly is the reverse order of disassembly. Clean mating surfaces of speed transmission and range transmission and apply a bead of Loctite 515 to face of speed transmission. Tighten speed housing to range housing bolts to a torque of 312 N·m (30 ft.-lbs.). Tighten cab mounting nuts to 312-380 N·m (230-280 ft.-lbs.) and rear wheel nuts to 434-475 N·m (320-350 ft.-lbs.).

FORWARD-REVERSE SHUTTLE VALVE

The forward-reverse shuttle valve is operated by a hand lever on the instrument panel. The lever operates an electrical switch that sends signals to the solenoid valves in the shuttle valve, which is mounted on top of the rear main frame. The solenoid valves control the pressurized fluid for operation of the forward and reverse clutches. Therefore, problems in the system can be electrical or hydraulic.

All Models

113. TROUBLESHOOTING. If forward-reverse unit is not operating properly, check as follows: Disconnect the forward-reverse switch (9—Fig. 136) connector from the wiring harness. With key switch in OFF position, move forward-reverse lever (3) to forward position. Use a multimeter and check for continuity between terminals 1 and 2, 4 and 5, and then 7 and 8 of connector on switch side. All three tests should show continuity. If not, renew the switch. Move forward-reverse lever to reverse position. Check for continuity between terminals 2 and 3 and terminals 5 and 6. Both tests should show continuity. If not, renew the switch. Reconnect wiring.

To check the forward solenoid (Fig. 137), with key switch OFF, disconnect the forward solenoid connector from the wiring harness. Connect the multimeter leads between the two terminals of the forward solenoid connector. Reading should be 8 ohms. If not, renew the forward solenoid.

Check the reverse solenoid in the same manner. If a reading of 8 ohms is not obtained, renew reverse solenoid.

The modulation solenoid can be checked using the same procedure. If a reading of 8 ohms is not obtained, renew the modulation solenoid.

The differential lock solenoid can be checked in the same manner as other solenoids. If a reading of 8 ohms is not obtained, renew differential lock solenoid.

To check the forward pressure switch, reconnect all wires to the control valve. Then, with the key switch OFF, disconnect the forward pressure switch connector from the wiring harness. Connect the leads of a multimeter between the two terminals of the pressure switch connector (switch side) and check for continuity. Test should show continuity. If not, renew the pressure switch. Then, start and operate engine at 1500 rpm. With range transmission in NEUTRAL and the forward-reverse control lever in forward position, cycle the inching (clutch) pedal. Check for continuity between the two terminals on switch side of connector. Test should not show continuity. If continuity is shown, renew forward pressure switch. Reconnect wiring.

To check the reverse pressure switch, with key switch OFF, disconnect the reverse pressure switch connector from the wiring harness. Connect the leads of a multimeter between the two terminals of the pressure switch connector (switch side) and check for continuity. Test should show continuity. If not, renew the pressure switch. Then, start engine and operate at 1500 rpm. With range transmission in NEUTRAL and the forward-reverse control lever in reverse position, cycle the inching pedal. Check for continuity between the two terminals on switch side of connector. Test should not show continuity. If continuity is shown, renew reverse pressure switch and reconnect wiring.

If the electrical components test good, check hydraulic pressures as follows: First, check the regulated pressure. Remove the test plug on synchromesh models, or Powershift supply line on Powershift models, from the top of the hydraulic pump compensator valve and install a 4000 kPa (600 psi) test gauge. Heat hydraulic oil to a temperature of 50° C (120° F). Operate engine at 1500 rpm and fully cycle the inching pedal. Pressure must be 1800-2070 kPa (260-300 psi). If not, renew the compensator. Remove test gauge and reinstall plug or Powershift supply line.

To test the forward and reverse pressures, refer to Fig. 137 and remove plugs from pressure ports (FP and RP). Install two 4000 kPa (600 psi) test gauges

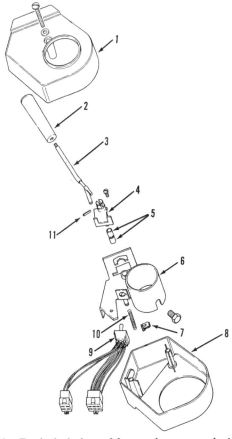

Fig. 136—Exploded view of forward-reverse shuttle control lever, switch and relative components.

1. Cover (upper)	
2. Knob	
3. Lever	7. Clip
4. Pivot	8. Cover (lower)
5. Bushings	9. Switch
6. Mounting bracket	10. Spring
	11. Roll pin

Fig. 137—View showing location of electrical components and pressure test ports of forward-reverse control valve.

FP. Forward pressure test port
RP. Reverse pressure test port

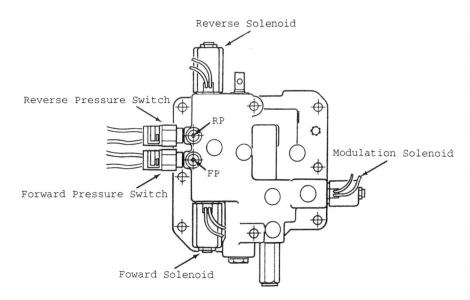

into the ports. Heat hydraulic oil to a temperature of 50° C (120° F) and operate engine at 1500 rpm. With range transmission in NEUTRAL and speed transmission in first speed position, fully cycle inching pedal. Move forward-reverse control lever to forward, then reverse positions and note pressure readings. Pressure readings should be 1800-2000 kPa (260-290 psi). If not, remove, clean and repair forward-reverse control valve assembly as outlined in paragraph 114.

114. R&R AND OVERHAUL. To remove the forward-reverse control valve, apply park brake or block rear wheels securely. Disconnect battery cables.

NOTE: Although not entirely necessary, some mechanics prefer to remove seat assembly to provide easier access for cleaning and removal of the valve.

Thoroughly clean control valve and surrounding area. Identify and tag pressure switch and solenoid electrical connectors, then disconnect the connectors. Disconnect the inching cable. Unbolt the chassis ground wire. Disconnect inching spool return spring. Remove the inching spool link. Refer to Fig. 138 and remove valve mounting bolts (6 and 12). Remove the shuttle valve and place on a clean bench.

To disassemble the forward-reverse shuttle valve, remove breather (15). Mark solenoids and valve body (1) for aid in reassembly. Identify and mark pressure switches, then remove the pressure switches (4 and 5). Remove the reverse solenoid (10) retaining nut and coil block, then unscrew and remove the solenoid valve. Repeat the operation for the forward solenoid (3), modulation solenoid (17) and the differential lock solenoid (18). Remove the temperature compensating valve (31), orifice spool (28) and filter (27).

NOTE: The temperature compensating valve is not serviceable and should be disassembled only for cleaning. Renew the valve if any fault is found.

Remove the front modulation valve plug, then the rear modulation valve plug (21) and remove the modulation piston (23), inner spring (24), outer spring (25) and spool (26). Remove front sequencing valve plug and rear sequencing valve plug (39), then remove spring (41), spacer (42) and sequencing spool (43). Remove the four Allen screws and remove cap (32), secondary spring (34), washer (35), return spring (36) and poppet (37). Withdraw the inching spool (38). Use a seal puller such as (CAS-10580) and remove oil seal (11).

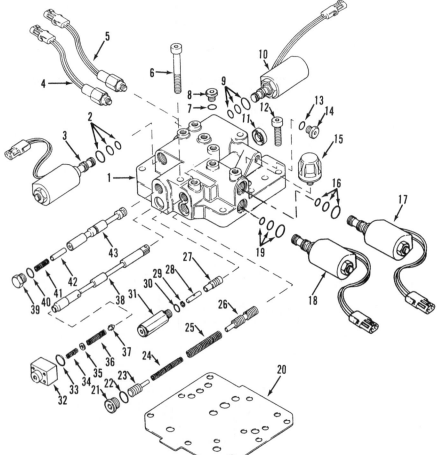

Fig. 138—Exploded view of the forward-reverse control valve assembly used on all models.

1. Valve body	23. Modulation piston
2. "O" rings	24. Modulation
3. Forward solenoid	spring (inner)
4. Forward pressure	25. Modulation
switch	spring (outer)
5. Reverse pressure	26. Modulation
switch	spool
6. Retaining bolt (2)	27. Filter
7. "O" ring	28. Orifice spool
8. Plug	29. Snap ring
9. "O" rings	30. "O" ring
10. Reverse solenoid	31. Temperature
11. Oil seal	compensator valve
12. Retaining bolt (6)	32. Cap
13. "O" ring	33. "O" ring
14. Plug	34. Secondary spring
15. Breather	35. Washer
16. "O" rings	36. Return spring
17. Modulation	37. Poppet
solenoid	38. Inching (clutch)
18. Differential	spool
lock solenoid	39. Plug
19. "O" rings	40. "O" ring
20. Gasket	41. Sequencing spring
21. Plug	42. Spacer
22. "O" ring	43. Sequencing spool

Clean and inspect all parts for excessive wear or other damage. Check springs against the following specifications. Refer to Fig. 138 for spring numbers.

Modulation spring (inner) #24:
Free length...................... 92.62 mm
(3.65 in.)

Modulation spring (outer) #25:
Free length...................... 97.81 mm
(3.85 in.)

Sequencing spool spring #41:
Free length...................... 59.28 mm
(2.334 in.)

Inching spool return spring #36:
Free length...................... 40.95 mm
(1.612 in.)

Inching spool secondary spring #34:
Free length...................... 19.3 mm
(0.76 in.)

When reassembling, use new oil seal and all new "O" rings. Reassemble in reverse order of disassembly. Tighten the forward solenoid valve (3) and the reverse solenoid valve (10) to a torque of 54-61 N·m (40-46 ft.-lbs.). Tighten the modulation solenoid valve (17) and the differential lock solenoid valve (18) to a torque of 16.3-19 N·m (12-14 ft.-lbs.). Tighten pressure port plugs to 8-14 N·m (6-10 ft.-lbs.).

Using a new gasket (20), reinstall the assembly by reversing the removal procedure. Tighten the shuttle valve retaining bolts (6 and 12) to a torque of 61 N·m (45 ft.-lbs.). Refer to paragraph 115 for adjustment of inching cable.

INCHING PEDAL AND TRANSMISSION LINKAGE ADJUSTMENT

All Models

115. INCHING PEDAL. To adjust the inching pedal, refer to Fig. 139 and disconnect inching cable from the forward-reverse shuttle valve spool. Make sure the inching cable ball socket (3) is threaded 12 mm (0.472 in.) onto the cable and that locknut is tight. Check to be sure that inching pedal is against stop "X" in the engaged position. Adjust the inching cable adjusting nuts (1) on cable support bracket (2) until the ball socket (3) on the cable can be freely installed onto the bellcrank ball stud. Lubricate cable end socket and ball joint with special grease (Part No. 3050944R1). Lock the adjusting nuts. This procedure will provide inching pedal travel, as measured at the center of the pedal pad, of 127 mm (5.0 in.). It will also provide 32.9 mm (1.295 in.) of pedal over travel.

116. INCHING PEDAL SWITCHES. To adjust the inching pedal switches (1 and 2—Fig. 140) on models equipped with standard transmissions (without creeper), proceed as follows: Make sure inching pedal is fully released and against pedal stop "X." Detach electrical connector from switch (1) and attach multimeter leads to terminals 1 and 2 of switch (1). Loosen the locknut and back switch out (counterclockwise) until there is continuity. Then, turn switch clockwise until there is no continuity. Remove the multimeter. Turn switch (1) carefully clockwise a minimum of ¼ turn to a maximum of 1¼ turns until terminals of switch face left side of tractor. Tighten locknut and attach wiring connector.

To adjust switch (2), the inching pedal must be held fully down and against pedal stop "Y." Detach electri-

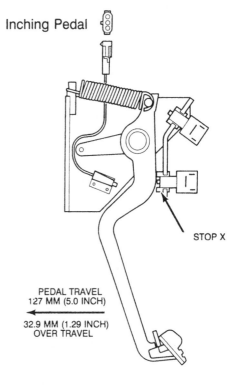

Inching Pedal

STOP X

PEDAL TRAVEL
127 MM (5.0 INCH)

32.9 MM (1.29 INCH)
OVER TRAVEL

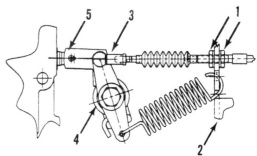

Fig. 139—View showing inching pedal and cable adjustment.
1. Cable adjusting nuts
2. Bracket
3. Ball socket
4. Bellcrank
5. Shuttle valve spool

cal connector from switch (2) and attach multimeter leads to terminals 3 and 4 of switch (2). Loosen the locknut and back switch out (counterclockwise) until there is no continuity. Then, turn switch clockwise until there is continuity. Remove the multimeter. Carefully turn switch (2) clockwise a minimum of ¼ turn to a maximum of 1¼ turns until terminals of switch face left side of tractor. Tighten locknut and attach wiring connector.

On models equipped with Powershift and creeper transmission up to and including P.I.N. No. JJF1005652, inching pedal switch (1) is the same as switch (1) on standard transmission models and adjustment is the same. However, switch (3—Fig. 141) is used in place of switch (2—Fig. 140). To adjust switch (3—Fig. 141), the inching pedal must be held fully down and against pedal stop "Y." Detach the electrical six-terminal connector from switch (3) and attach multimeter leads to terminals 5 and 6 of switch (3). Loosen locknut and turn switch counterclockwise until there is no continuity. Then, turn switch clockwise until there is continuity between terminals 5 and 6. There also will be continuity

between terminals 3 and 4, but no continuity between terminals 1 and 2. Detach the multimeter. Carefully turn switch (3) clockwise a minimum of ¼ turn to a maximum of 1¼ turns until terminals of switch face left side of tractor. Tighten locknut and connect the wiring connector.

On models equipped with Powershift or Powershift and creeper transmission (P.I.N. No. JJF1005653 and later), inching pedal switches (1 and 2) are the same as switches (1 and 2—Fig. 140) used on standard transmission models and adjustment of these switches is the same. However, switch (4—Fig. 142) is also used on these models. To adjust switch (4), inching pedal must be held fully down and against pedal stop "Y." Detach electrical connector from switch wires and attach multimeter leads to switch terminals B and C. Loosen the switch support screws and move switch support down until continuity is obtained. Then, continue to move support bracket down 1 mm (0.039 in.) and tighten support screws. Connect wiring connector.

117. SYNCHROMESH SPEED AND RANGE TRANSMISSION LINKAGE. Adjustment of the speed and range transmission linkage is similar. Move levers on side of transmission to position shift rails in neutral detent position. Refer to Fig. 143 and loosen locknuts on control rods. Adjust length of control rods so that selector lever can be moved from the 1st/2nd side to the 3rd/4th side without binding or hanging up. Lubricate ends of linkage rods with special grease such as (Part No. 3050944R1). Tighten locknuts.

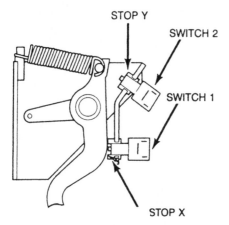

Fig. 140—View of inching pedal switches used on models with standard transmission (without creeper).

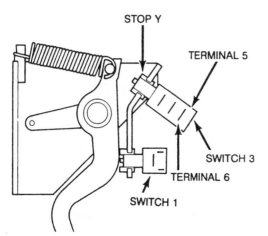

Fig. 141—View of inching pedal switches used on models with Powershift and creeper transmission up to and including P.I.N No. JJF1005652.

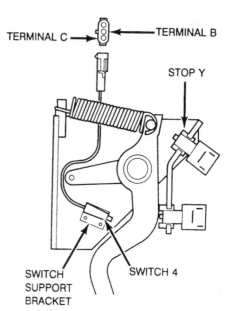

Fig. 142—View of inching pedal switches used on models with Powershift or Powershift and creeper transmission (P.I.N. No. JJF1005653 and later).

118. RANGE TRANSMISSION SENSING SWITCHES.

Before adjusting transmission sensing switches, make certain shifting linkage is adjusted as in paragraph 117. Disconnect wiring connectors at transmission sensing switches (Fig. 144). Identify and tag all connectors for aid in reassembly.

NOTE: If necessary to renew switches, apply a drop of Loctite 242 to the switch threads.

To adjust and test the 1st/2nd range sensing switch (1—Fig. 145), shift 1st/2nd shift lever to 1st range position. Attach multimeter leads to terminals (C and D) of switch connector. Loosen the jam nut and turn switch (1) counterclockwise until multimeter shows no continuity. Then, turn switch clockwise until continuity is shown. Carefully turn the switch clockwise an additional ½ turn. Tighten jam nut to a torque of 35-40 N·m (26-30 ft.-lbs.). Using the chart shown in Fig. 145, test the 1st/2nd switch (1). If unable to obtain continuity as shown, renew the switch.

To adjust and test 3rd/4th range sensing switch (2—Fig. 146), move 3rd/4th shift lever to 3rd range position. Attach multimeter leads to terminals (C and D) of switch connector. Loosen the jam nut and turn switch (2) counterclockwise until multimeter shows no continuity. Then, turn switch clockwise until continuity is shown. Carefully turn switch clockwise an

additional ½ turn. Tighten jam nut to a torque of 35-40 N·m (26-30 ft.-lbs.). Using the chart shown in Fig. 146, complete the tests. If unable to obtain continuity as shown, renew the switch.

To adjust and test the 4th range sensing switch (3—Fig. 147), shift 3rd/4th shift lever to 3rd range position. Attach multimeter leads to terminals (C and D) of switch connector. Loosen jam nut and turn switch (3) counterclockwise until multimeter shows no continuity. Then, turn switch clockwise until con-

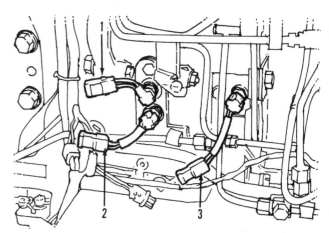

Fig. 144—Disconnect range sensing switch connectors at locations (1, 2 and 3).

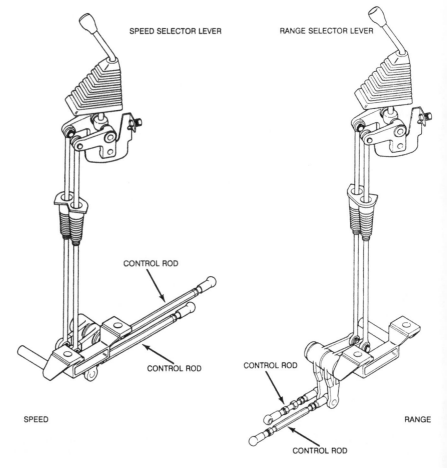

Fig. 143—View showing control rods and selector lever linkage for synchromesh speed and range transmissions.

SPEED SELECTOR LEVER

RANGE SELECTOR LEVER

CONTROL ROD

CONTROL ROD

CONTROL ROD

CONTROL ROD

SPEED

RANGE

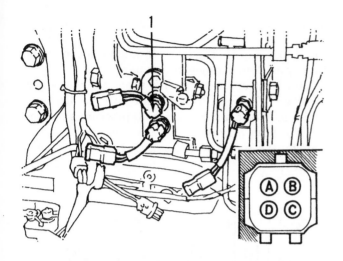

Range Selection	Terminals	Continuity Reading
1st Range	D and C	Yes
1st Range	A and B	No
Neutral	D and C	No
Neutral	A and B	Yes
2nd Range	A and B	No
2nd Range	D and C	Yes

Fig. 145—View of 1st and 2nd range switch (1), switch connector terminals and test chart.

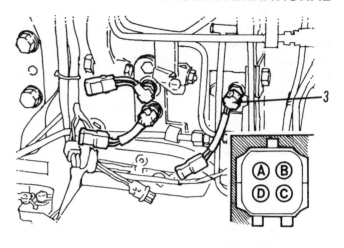

Range Selection	Terminals	Continuity Reading
3rd Range	D and C	Yes
3rd Range	A and B	No
Neutral	D and C	Yes
Neutral	A and B	No
4th Range	A and B	Yes
4th Range	D and C	No

Fig. 147—View of 4th range sensing switch (3), switch connector terminals and test chart.

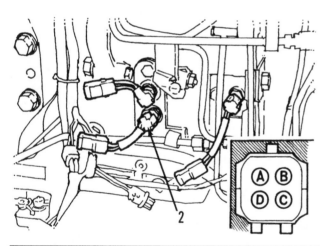

Range Selection	Terminals	Continuity Reading
3rd Range	D and C	Yes
3rd Range	A and B	No
Neutral	D and C	No
Neutral	A and B	Yes
4th Range	A and B	No
4th Range	D and C	Yes

Fig. 146—View of 3rd/4th range sensing switch (2), switch connector terminals and test chart.

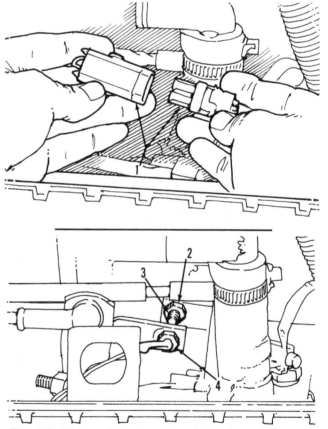

Fig. 148—Upper view shows synchromesh neutral sensing switch connector (1). Lower view shows jam nut (2), magnet (3) and reed switch (4) of the synchromesh neutral sensing switch.

tinuity is shown. Carefully turn switch clockwise an additional ½ turn. Tighten jam nut to a torque of 35-40 N·m (26-30 ft.-lbs.). Using the chart shown in Fig. 147, complete the tests. If unable to obtain continuity as shown, renew the switch.

119. SYNCHROMESH SPEED NEUTRAL SENSING SWITCH. To adjust and test the synchromesh speed neutral switch, refer to Fig. 148 and disconnect switch connector (1). With speed selector lever in neutral position, loosen magnet jam nut (2). Turn magnet (3) clockwise until it touches the reed switch (4). Then, turn the magnet counterclockwise two turns and tighten the jam nut (2) to a torque of 3.4 N·m (2.5 ft.-lbs.). Attach multimeter leads to terminals (A and B—Fig. 149) of switch connector. Using the chart shown, check the continuity readings. If unable to obtain correct continuity readings, renew the switch.

NOTE: If switch is being renewed, apply a drop of Loctite 242 to threads of switch and magnet.

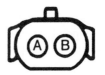

Speed Selection	Terminals	Continuity Reading
Neutral	A and B	Yes
1st	A and B	No
2nd	A and B	No
3rd	A and B	No
4th	A and B	No

Fig. 149—View of connector terminals and test chart for synchromesh neutral sensing switch.

DRIVE SHAFT
(WITHOUT CREEPER)

A standard drive shaft (Fig. 150) is used on all models without creeper transmission to connect the speed transmission to the range transmission.

All Models So Equipped

120. R&R AND OVERHAUL. To remove the standard drive shaft, engage park brake and block front wheels securely. Disconnect battery cables. Support axle housing and remove right rear wheel. Remove plugs and drain transmission oil. On synchromesh models, disconnect transmission shift control rods. Unbolt and remove access cover (Fig. 151) from right side of speed transmission. Working through the opening, remove bolt (17—Fig. 150), washer (18), shim (19) and drive gear (20) from rear of speed transmission drop

shaft. Bend back lock tabs on lockwasher (6). Remove snap ring (8) from its groove and slide forward on shaft. Use special tool (CAS-2016) to remove nut (7). Slide nut, lockwasher (6), thrust washer (5), bearing cone (4) and collar (1) forward on shaft. Then, remove assembly through the access opening. Remove collar (1), bearing cone (4), thrust washer (5), lockwasher (6), nut (7) and snap ring (8) from rear of drive shaft (9). Use a suitable puller and remove bearing cone (12). Remove driven gear (11).

Clean and inspect all parts and renew any showing excessive wear or damage. If bearing cones (4 and 12) are renewed, install new bearing cups (3 and 13). Remove snap ring (14), shim (15) and oil slinger (16). Install new oil slinger, then install shims until snap ring can just be installed. Shims (15) are available in thicknesses of 0.1, 0.356 and 0.5 mm (0.004, 0.014 and 0.020 in.). Heat bearing cones (4 and 12) in bearing oven to 150° C (302° F) during installation.

Reassemble by reversing disassembly procedure. Install assembly in housing. Slide collar (1) rearward into housing and move shaft assembly forward until bearing cone (12) is seated in bearing cup (13). Thread nut (7) onto collar (1). Install snap ring (8) in its groove in shaft (9). Using a feeler gauge, measure gap between snap ring and end of collar. Gap should be 0.076 mm (0.003 in.). Using special wrench (CAS-2016), adjust nut until correct measurement is obtained. When end play is correct, bend one tab of lockwasher into a groove on nut. Install drive gear (20), bolt (17), washer (18) and shims (19). Tighten bolt to 101-113 N•m (75-83 ft.-lbs.).

Clean mating surfaces of transmission housing and access cover. Apply a bead of Loctite 515 to transmission housing. Install access cover and tighten bolts securely. Connect synchromesh transmission shift control rods. Reinstall rear wheel, fill transmission to dipstick level with Hy-Tran Plus oil and reconnect battery cables.

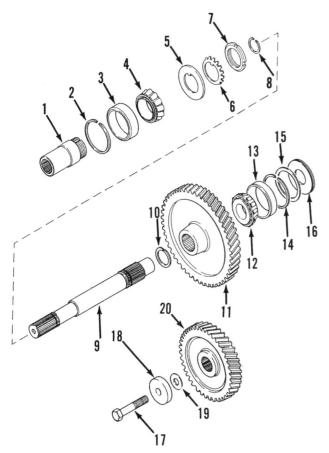

Fig. 150—Standard drive shaft and relative components used on all models not equipped with creeper transmission.

1. Collar		
2. Snap ring	8. Snap ring	14. Snap ring
3. Bearing cup	9. Drive shaft	15. Shim
4. Bearing cone	10. Snap ring	16. Oil slinger
(rear)	11. Driven gear	17. Bolt
5. Thrust washer	12. Bearing cone	18. Washer
6. Lockwasher	(front)	19. Shim
7. Nut	13. Bearing cup	20. Drive gear

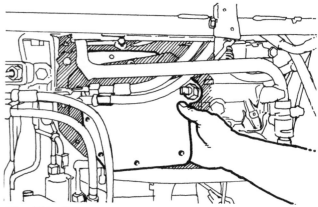

Fig. 151—Remove access cover from right side of speed transmission.

CREEPER TRANSMISSION

A creeper transmission is optionally available and is located directly behind the speed transmission. The creeper transmission can be used for pto-type harvesting and other operations where slow ground speeds are desirable. The creeper drive provides tractor with 8 low speeds forward and 8 low speeds in reverse. Ranges 3 and 4 are locked out for creeper operation. The creeper transmission is used in place of the main drive shaft outlined in paragraph 120.

All Models So Equipped

121. R&R AND OVERHAUL. To remove the creeper transmission, engage park brake and block front wheels securely. Disconnect battery cables. Support axle housing and remove right rear wheel. Remove plugs and drain transmission oil. On synchromesh models, disconnect and remove speed transmission shift control rods. Disconnect creeper shift cable from shift lever (11—Fig. 152). Loosen adjusting nuts and remove control cable from

bracket. Disconnect creeper lubrication tube. Place index marks on creeper shift lever (11—Fig. 152) and end of shifter shaft (9). Loosen clamping bolt (12) and remove shift lever (11). Unbolt and remove creeper housing assembly.

To disassemble the creeper housing assembly, refer to Fig. 153 and remove snap rings (1 and 7). Using a bushing driver, drive reduction shaft (4) rearward as far as possible. Then, drive the shaft forward as far as possible. Use a bearing cup puller and slide hammer to pull bearing cups (2 and 6) from the housing. Using a suitable puller, remove rear bearing cone (3) from the shaft. Remove reduction shaft from housing. Remove front bearing cone (5) from the shaft. Remove the shifter shaft oil seal (10—Fig. 152) from creeper housing.

Clean and inspect all parts for excessive wear or other damage and renew as necessary.

To reassemble, heat front bearing cone (5—Fig. 153) to a temperature of 150° C (302° F) and install on shaft until fully seated. Install bearing cup (6) into housing, then install the thinnest snap ring (7) that is available.

NOTE: Snap rings (1 and 7) are available in thicknesses of 1.5, 1.6, 1.7, 1.8 and 2.0 mm (0.059, 0.063, 0.067, 0.071 and 0.079 in.).

Install reduction shaft (4) into housing (8). Heat rear bearing cone (3) to 150° C (302° F) and install on shaft until fully seated. Install rear bearing cup (2) into housing until a 1.5 mm (0.059 in.) thick snap ring (1) can be installed. Drive reduction shaft forward

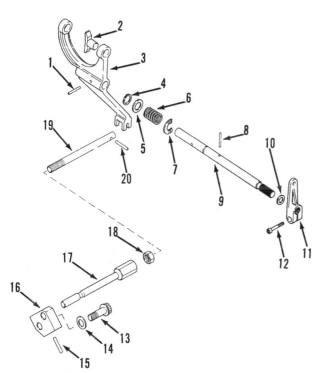

Fig. 152—Exploded view of creeper transmission internal shift linkage.

1. Groove pin
2. Shift finger (2)
3. Shift fork
4. Snap ring
5. Washer
6. Spring
7. Detent washer
8. Pin
9. Shifter shaft
10. Oil seal
11. Shift lever
12. Clamping bolt
13. Bolt
14. Washer
15. Interlock pin
16. Interlock housing
17. Interlock rod (rear)
18. Nut
19. Interlock rod (front)
20. Roll pin

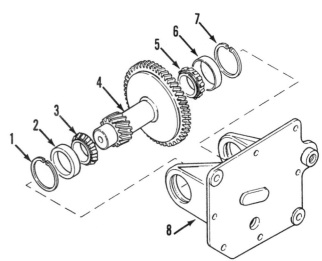

Fig. 153—View of reduction shaft removed from creeper transmission cover.

1. Snap ring
2. Bearing cup
3. Bearing cone (rear)
4. Reduction shaft
5. Bearing cone (front)
6. Bearing cup
7. Snap ring
8. Creeper cover

and rearward to seat bearing cups (2 and 6) against snap rings (1 and 7). Use a dial indicator to measure reduction shaft end play. End play should be 0.025-0.127 mm (.001-0.005 in.). Install different thickness snap rings (1 and 7) until end play is correct. Install a new shifter shaft oil seal in the housing (8). Lay assembly aside for later installation.

Refer to Fig. 152 and remove shifter shaft and fork assembly (1 through 9). Drive out groove pin (1) and remove shift fork (3). Remove snap ring (4), washer (5), spring (6) and detent washer (7) from shifter shaft (9).

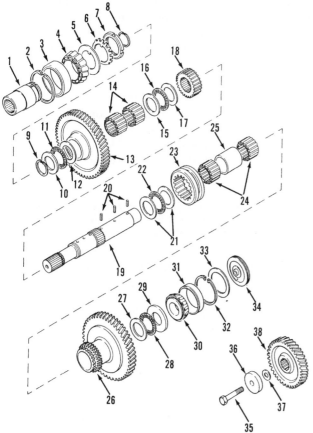

Fig. 154—Exploded view of creeper transmission drive shaft and gears.

1. Collar
2. Snap ring
3. Bearing cup
4. Bearing cone (rear)
5. Thrust washer
6. Lockwasher
7. Nut
8. Snap ring
9. Snap ring
10. Thrust washer
11. Thrust bearing
12. Thrust washer
13. Low speed drive gear
14. Roller bearings
15. Thrust washer
16. Thrust bearing
17. Thrust washer
18. Shift hub
19. Creeper drive shaft
20. Roll pins
21. Thrust washers
22. Thrust bearing
23. Shift collar
24. Roller bearings
25. Spacer
26. Direct drive gear
27. Thrust washer
28. Thrust bearing
29. Thrust washer
30. Bearing cone (front)
31. Bearing cup
32. Snap ring
33. Shim
34. Oil slinger
35. Bolt
36. Washer
37. Shim
38. Drive gear

Clean and inspect all parts and renew any showing excessive wear or other damage. Replacement shift fingers (2) are available for installation. Reassemble in reverse order of disassembly.

To remove the creeper drive shaft and gears from the transmission housing, refer to Fig. 154 and remove bolt (35), washer (36), shims (37) and drive gear (38) from rear of speed transmission drop shaft. Bend back the lock tab on lockwasher (6). Remove snap ring (8) from its groove and slide it forward on shaft. Use special tool (CAS-2016) and remove nut (7). Slide nut, lockwasher (6), thrust washer (5), bearing cone (4) and collar (1) forward on shaft. Then, remove creeper shaft and gear assembly from transmission housing. Remove collar (1), bearing cone (4), thrust washer (5), lockwasher (6), nut (7) and snap ring (8) from rear of creeper drive shaft (19). Remove snap ring (9), thrust washers (10 and 12), thrust bearing (11) and low speed drive gear (13). Remove roller bearings (14), thrust washers (15 and 17), thrust bearing (16), shift collar (23) and shift hub (18). Using a suitable puller, remove front bearing cone (30). Remove thrust washer (29), thrust bearing (28), thrust washer (27) and direct drive gear (26). Remove roller bearings (24), spacer (25), thrust washers (21) and thrust bearing (22).

Clean and inspect all parts and renew any showing excessive wear or other damage. If bearing cones (4 and 30) are to be renewed, renew bearing cups (3 and 31).

Remove snap ring (32), shim (33) and oil slinger (34). Install a new oil slinger, then install shims (33) until snap ring can just be installed. Shims (33) are available in thicknesses of 0.1, 0.356 and 0.5 mm (0.004, 0.014 and 0.020 in.). Heat bearing cones (4 and 30) in a bearing oven to a temperature of 150° C (302° F) during installation.

Reassemble by reversing the disassembly procedure. Install creeper shaft and gear assembly into the transmission housing. Slide collar (1) rearward into the housing and move shaft forward until bearing cone (30) is seated in bearing cup (31). Thread nut (7) onto collar (1). Install snap ring (8) in its groove in shaft (19). Using a feeler gauge, measure gap between snap ring and end of collar. This gap should be 0.076 mm (0.003 in.). Using special wrench (CAS-2016), adjust nut until correct measurement is obtained. When the end play is correct, bend one of the tabs of lockwasher (6) into groove on nut. Install drive gear (38), bolt (35), washer (36) and shims (37). Tighten bolt to a torque of 101-113 N•m (75-83 ft.-lbs.).

Install shift fork (3—Fig. 152) and shifter shaft (9) assembly, making sure shift fingers (2) are engaged in the shift collar and lower end of fork engages pin (20) in interlock rod (19).

Clean mating surfaces of transmission housing and creeper housing (8—Fig. 153). Apply a bead of Loctite 515 to transmission housing. Install creeper housing assembly and tighten bolts securely. Align index marks on shift lever (11—Fig. 152) and end of shifter shaft (9), install shift lever and tighten clamping bolt (12). Connect synchromesh transmission shift control rods.

To adjust the creeper shift cable (7—Fig. 155), move shift lever (1) to disengaged position. With clevis (5) disconnected from creeper transmission lever (11—Fig. 152), move the lever to disengaged position. Loosen adjusting nuts (6—Fig. 155) and adjust cable (7) until clevis (5) aligns with hole in transmission lever. Install clevis pin and tighten nuts (6) against support bracket (4). To adjust the creeper warning light switch (2), move shift lever and check operation of the switch. After initial contact, the switch plunger should move 10 mm (0.4 in.). Adjust position of switch to give correct movement, then tighten locknut (7).

Reinstall rear wheel, fill transmission with Hy-Tran Plus oil to dipstick level and reconnect battery cables.

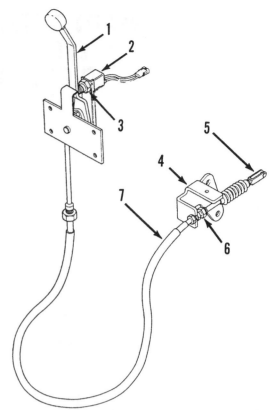

Fig. 155—View of creeper transmission shift lever and cable assembly. Refer to text for adjustment procedure.

1. Shift lever
2. Warning light switch
3. Locknut
4. Bracket
5. Clevis
6. Adjusting nuts
7. Control cable

MAIN DRIVE BEVEL GEARS, DIFFERENTIAL AND DIFFERENTIAL LOCK

DIFFERENTIAL AND BEVEL GEARS

All Models

122. To remove the differential assembly, first remove the final drive assemblies as outlined in paragraph 124. Remove the pto housing and pto drive shaft as outlined in paragraph 166. Remove outer brake backing plate and friction plate from each side, then remove sun gear shafts (1 and 43—Fig. 156). Disconnect and remove left bearing carrier lubrication tube (12). Support differential with wooden blocks. Remove bearing carrier retaining bolts (42) and install two alignment studs. Use two bolts (42) as jack screws in the threaded holes in carrier. Tighten bolts evenly until bearing carrier can be removed. Remove shims (39) with carrier (40). Remove left side bearing carrier (5) in the same manner. Remove differential assembly and oil seal (6) from rear main frame. Remove seal ring (2) and bearing cone (7) from left carrier (5) and remove seal rings (33), "O" rings (38) and bearing cone (37) from right carrier (40).

To disassemble the differential, first place alignment marks across differential lock cage (31), differential cage (11) and ring gear (13). Unbolt and remove cage (31). Remove reaction plate (23), hub (24), friction discs (25), separator plates (26) and piston return plate (27). Bump cage (31) against a wood block to remover piston (28). Remove seal rings (29 and 30) from piston. Remove side gear (21), then unbolt and remove ring gear (13). Place index marks on cage (11) and on end of differential pins (17 and 20). Drive out both roll pins (22) and remove pins (17 and 20), differential pinions (16), thrust washers (15), left side gear (19) and thrust washer (18).

Clean and inspect all parts and renew any showing excessive wear or other damage. Thickness of new friction disc is 2.13 mm (0.084 in.) with a wear limit of 1.98 mm (0.078 in.). Thickness of new separator plate is 1.18 mm (0.046 in.) with a wear limit of 1.05 mm (0.041 in.). Thickness of new piston return plate is 4.23 mm (0.167 in.) with a wear limit of 4.11 mm (0.162 in.). The ring gear (13) and drive pinion shaft (14) are a matched set and cannot be serviced separately.

If drive pinion shaft is to be renewed, refer to RANGE TRANSMISSION paragraph 112 for procedure. If bearing cones (7 and 37—Fig. 156) are to be renewed, install new bearing cups (8 and 36) in car-

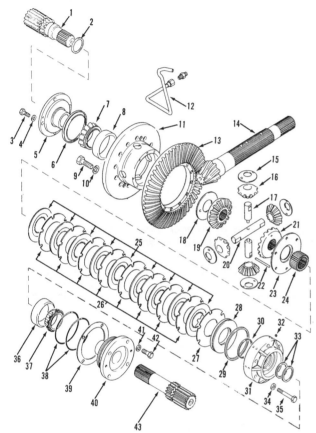

Fig. 156—Exploded view of main drive bevel gears, differential and differential lock assembly.

1. Sun gear shaft (L.H.)
2. Seal ring
3. Bolt
4. Hardened washer
5. Bearing carrier (L.H.)
6. Oil seal
7. Bearing cone
8. Bearing cup
9. Bolt
10. Hardened washer
11. Differential cage
12. Differential lube tube
13. Ring gear
14. Drive pinion shaft
15. Thrust washer
16. Differential pinions (4)
17. Differential pin (short)
18. Thrust washer
19. Side gear (left)
20. Differential pin (long)
21. Side gear (right)
22. Roll pin (2)
23. Reaction plate
24. Splined hub
25. Friction discs
26. Separator plates
27. Piston return plate
28. Piston
29. Seal ring (outer)
30. Seal ring (inner)
31. Differential lock cage
32. Ball plug
33. Seal rings
34. Hardened washer
35. Bolt
36. Bearing cup
37. Bearing cone
38. "O" rings
39. Shim
40. Bearing carrier (R.H.)
41. Hardened washer
42. Bolt
43. Sun gear shaft (R.H.)

riers (5 and 40). Install thrust washer (18) and side gear (19) in cage (11). Use petroleum jelly to hold thrust washers (15) in place and install differential pinions (16). Install long pin (20), then short pins (17) and secure with roll pins (22). Install two guide studs into the ring gear. Heat ring gear in a bearing oven to a temperature of 150° C (302° F), then install ring gear on differential cage (11). If original parts are used, align the marks applied during disassembly. Install retaining bolts (9) with new hardened washers (10) and tighten bolts to a torque of 265 N·m (195 ft.-lbs.). Install side gear (21) and install two 10 mm guide studs into the differential cage. Install reaction plate (23) with flat side away from side gear (22). Starting with a friction disc (25) followed by a separator plate (26), alternately install the nine friction discs and eight separator plates. Install piston return plate (27). Lubricate inner and outer seal rings (29 and 30) with petroleum jelly and install on piston (28). Install piston assembly into the differential lock cage (31). Align assembly marks and install cage (31) onto cage (11). Remove alignment studs and install bolts (35) with new hardened washers (34). Tighten bolts to a torque of 62 N·m (46 ft.-lbs.). Heat bearing cones (7 and 37) and install on bearing carriers (5 and 40) until seated against shoulder. Lubricate seal ring (2) and install in left carrier (5). Install the differential assembly into rear main frame and support with wood blocks to align for installation of bearing carriers. Install two 10 mm guide studs in rear main frame, then install right side carrier (40) without seal rings (33 and 38) and shims (39). Install bolts (42) and new hardened washers (41) and tighten bolts to a torque of 54-61 N·m (40-45 ft.-lbs.). Install left side carrier (5) without seal (6) and shims, and tighten retaining bolts (3) to a torque of 20 N·m (15 ft.-lbs.). Remove wood support blocks. Rotate differential assembly and recheck torques. Loosen the left side mounting bolts (3), then, while rotating the differential, retorque the bolts to 9 N·m (6.6 ft.-lbs.).

Using a depth gauge, measure the distance between face of bearing carrier (5) and rear main frame through the two threaded holes. Average the two measurements and record. Block up under differential assembly and remove left bearing carrier (5). Using a micrometer, measure the thickness of bearing carrier flange. Subtract the flange thickness from the previously recorded averaged measurement. Then, add 0.114 mm (0.0045 in.) to determine the required shim pack thickness. Install the left bearing carrier with the determined shim pack. Shims are available in thicknesses of 0.08, 0.18, 0.25 and 0.63 mm (0.003, 0.007, 0.010 and 0.025 in.). Tighten bolts (3) to a torque of 77-87 N·m (57-64 ft.-lbs.), then tighten right carrier bolts (42) to 77-87 N·m (57-64 ft.-lbs.). Remove support blocks. Using a dial indica-

tor, measure backlash between ring gear (13) and drive pinion shaft (14). Backlash should be 0.15-0.30 mm (0.006-0.012 in.). To reduce backlash, move shims from left side to right side. For each 0.003 inch (0.08 mm) of shim thickness moved, backlash will change approximately 0.05 mm (0.002 in.).

When correct backlash between ring gear and pinion shaft has been set, support differential with wood blocks. Unbolt and remove left side bearing carrier (5) and shims. Install oil seal (6) in rear main frame so that seal is even with chamfer in housing. The seal will be moved to correct position when bearing carrier and shim pack are reinstalled. Install the bearing carrier and shim pack and tighten bolts to a torque of 77-87 N·m (57-64 ft.-lbs.). Remove right bearing carrier (40) and shims (39). Lubricate "O" rings (38) and seal rings (33) and install on bearing carrier. Install bearing carrier and shim pack and tighten bolts to a torque of 77-87 N·m (57-64 ft.-lbs.). Remove wood support blocks. Install left side bearing carrier lube tube (12).

Install pto drive shaft and pto housing as outlined in paragraph 166. Install the brake friction plate and outer brake backing plate on each side, then install sun gear shafts (1 and 43—Fig. 156). Install the final drive assemblies as outlined in paragraph 124. Fill transmission to FULL mark on dipstick with Hy-Tran Plus oil.

DIFFERENTIAL LOCK

All Models

123. All tractors are equipped with a differential lock that will make both rear wheels rotate at the same speed even if one wheel has little or no traction. The differential lock clutch pack and actuating piston are located within the differential and lock clutch housing. When operator depresses differential lock switch, an electric solenoid-controlled, hydraulic valve directs pressurized oil to actuate clutch assembly. The clutch locks the right differential side gear (21—Fig. 156) to the differential lock cage (31), which prevents differential action and causes both final drive sun gear shafts (1 and 43) to rotate at the same speed. When switch is turned off or foot brake is applied (automatic switch activated), actuating piston hydraulic pressure is relieved and clutch pack disengages.

Removal and installation of clutch pack and piston is outlined in paragraph 122 covering differential service. Solenoid control valve is located in the forward-reverse control valve (Fig. 138) and is removed and installed as outlined in paragraph 114. Refer to paragraph 113 to check the differential lock solenoid.

FINAL DRIVES

The final drive assemblies used on all models consist of rear axle, planetary reduction unit and axle housing. The final drive is removed from tractor as a unit.

All Models

124. REMOVE AND REINSTALL. To remove the final drive, first block front wheels securely and place wooden wedges between front axle and front support to prevent tipping. Remove plugs and drain transmission oil. Remove the lower hitch links from rear axle. Place a tractor stand (CAS-10853 or equivalent) under transmission and rear frame and raise rear of tractor. Remove rear wheels. Disconnect cab ground cables at each side. Remove rear cab mounting nuts and washers. Loosen cab front mounting bolts. Raise the hood and remove exhaust extension pipe and rear hood panel. Using cab support stands (CAS-3389), raise rear of cab to clear rear mounting bolts. Unbolt and remove rear cab mounting brackets. Remove two rear axle housing mounting bolts opposite each other and install alignment studs. Remove and cap the sensing tube. Support the axle assembly with a floor jack and remove the remaining bolts. Push downward on outer end of axle to balance the assembly, then remove final drive assembly from tractor.

Clean the rear axle housing and rear main frame mating surfaces and apply a bead of B500642 sealant to face of rear main frame. Apply activator B500622 to axle housing. Wait five minutes before continuing installation. Install rear axle housing on the alignment studs until the mounting faces are together. Install and tighten retaining bolts to a torque of 325-353 N·m (240-260 ft.-lbs.). Install the sensing tube. Install cab mounting brackets and tighten the

bolts to 310-380 N·m (230-280 ft.-lbs.). Carefully lower the cab and tighten the mounting bolts and nuts to a torque of 310-380 N·m (230-280 ft.-lbs.). Install rear wheels and on models with flanged rear axles, tighten wheel nuts to a torque of 434-475 N·m (320-350 ft.-lbs.). On models with rowcrop axles, tighten wheel hub bolts in four stages using the sequence shown in Fig. 157. Tighten bolts in first stage to 74 N·m (55 ft.-lbs.), in second stage to 150 N·m (110 ft.-lbs.), in third stage to 224 N·m (165 ft.-lbs.) and in the final stage to 300 N·m (220 ft.-lbs.). The balance of installation is the reverse of removal procedure. Fill transmission with Hy-Tran Plus oil to the full mark on dipstick. Bleed air from brake system as in paragraph 130.

125. OVERHAUL. To disassemble the final drive unit, stand the assembly upright on the axle flange (standard axle) or the tapered hub (rowcrop axle). Remove the retaining plate snap rings (1—Fig. 158), then remove retaining plate (2) and "D" washers (3). Place index marks or numbers on planetary gears and pins so they can be installed in correct positions. Use grease on screwdriver tip to remove top layer of needle rollers (4). Remove planetary gear (6), spacer (5) and second layer of needle rollers (4). Remove the other three planetary gears and needle rollers in the same manner. Remove thrust washers (7). Remove snap ring (8), then lift planetary carrier (9) from end of axle shaft. Bend lockwasher tab away from nut (10). Using special locknut socket (CAS-1997), remove nut from end of axle shaft. Remove lockwasher (11) and thrust washer (12). Lay axle assembly down and support axle housing on wood blocks. Use a soft hammer to drive axle shaft (22 or 24) from the housing (17). Remove inner bearing cone (13) as axle shaft is removed. Use a suitable puller and remove outer bearing cone (21) from axle shaft. Remove and discard oil seal (25).

Clean and inspect all parts and renew any showing excessive wear or other damage. If bearing cones (13 and 21) are to be renewed, install new bearing cups (14 and 20) in axle housing (17).

On models with flanged standard axle (22), lubricate and install new oil seal (25) on axle shaft so seal manufacturer's name and part number are toward outer end of axle. Do not install oil seal on rowcrop axle at this time. Heat outer bearing cones (21) to a temperature of 150° C (302° F) and install onto axle shaft until seated against shoulder. On models with flanged standard axle, refer to Fig. 159 and clamp seal installer tool (CAS-2004) around new oil seal (25—Fig. 158). Install axle shaft into axle housing. Heat inner axle bearing cone (13) to 150° C (302° F) and install on axle shaft. Install and tighten nut (10)

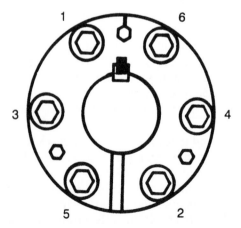

Fig. 157—On models equipped with rowcrop rear axles, use the sequence shown when tightening wheel hub bolts. Refer to text.

until oil seal installer tool contacts axle housing. Remove seal installer. Remove the nut, then install thrust washer (12), lockwasher (11) and nut (10). Tighten nut by hand. Stand axle assembly on flanged end of axle. Tighten nut with special socket (CAS-1997) until nut is tight to seat the bearings. Loosen the nut and hit splined end of axle shaft with a soft hammer to remove all bearing preload. Attach special tool as shown in Fig. 160 to check axle bearing rolling torque. Special tool can be fabricated in your shop using dimensions shown in Fig. 161. Using a torque wrench, first measure rolling torque with no bearing preload. Then, tighten nut (10—Fig. 158) until rolling torque increases 3.4-5.7 N•m (30-50 in.-lbs.) more

than rolling torque without bearing preload. When bearing preload is correct, bend tab on lockwasher (11) into slot on nut (10).

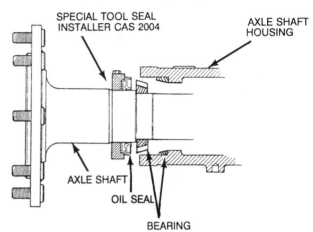

Fig. 159—Seal installer tool (CAS-2004) clamped around new seal to be installed in axle housing with flanged standard axle.

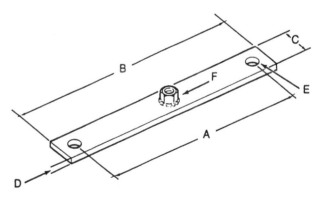

Fig. 160—Special tool attached to axle housing to check bearing preload.

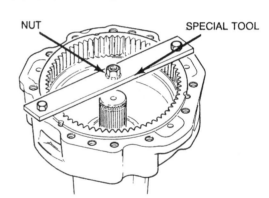

Fig. 158—Exploded view of final drive assembly used on all models.

1. Snap ring	14. Bearing cup
2. Retaining plate	15. Gear lock pin
3. "D" washer	16. Ring gear
4. Needle rollers	17. Axle housing
5. Spacer	18. Washer
6. Planetary gear	19. Bolt
7. Thrust washer	20. Bearing cup
8. Snap ring	21. Bearing cone (outer)
9. Planetary carrier	22. Axle shaft (standard)
10. Nut	23. Lug bolt
11. Lockwasher	24. Axle shaft (rowcrop)
12. Thrust washer	25. Oil seal
13. Bearing cone (inner)	

AXLE BEARING PRE-LOAD SETTING TOOL

A. 344 MM (13.5 INCH)	D. 5 MM (0.2 INCH)
B. 382 MM (15.0 INCH)	E. 20 MM (0.8 INCH)
C. 40 MM (1.6 INCH)	F. 14 MM (0.6 INCH)

Fig. 161—Special tool can be fabricated using the dimensions shown.

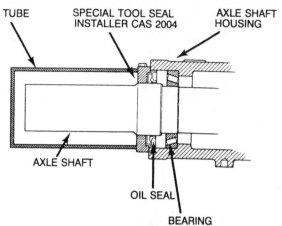

TUBE   SPECIAL TOOL SEAL   AXLE SHAFT
        INSTALLER CAS 2004   HOUSING

AXLE SHAFT

OIL SEAL

BEARING

Fig. 162—View showing axle oil seal installation on models equipped with rowcrop axles.

On rowcrop axles, install axle and bearings and set preload in same manner as the flanged axle. Then, lubricate and install oil seal (25) on axle shaft (24) so seal manufacturer's name and part number are toward outer end of axle. Clamp seal installer tool (CAS-2004) on seal, then drive the seal into axle housing using a tube over axle shaft as shown in Fig. 162.

On all models, install planetary carrier (9—Fig. 158) and secure with snap ring (8). Install thrust washer (7) and apply a coat of petroleum jelly to the washer and planetary pin. Carefully install lower layer of needle rollers (4). Install spacer (5) and matching index marks, install planetary gear (6). Install top layer of needle rollers (4) and lubricate with clean Hy-Tran oil. install the other three planetary gears in the same manner. Install "D" washers (3), retaining plate (2) and snap rings (1). Make sure the eyes of snap rings are opposite the flat on planetary pins.

BRAKES

Brakes on all models are hydraulically actuated, wet-type, single-disc-type. Brakes are located on the differential output shafts (planetary sun gear shafts) and are accessible after removing final drive units. Return hydraulic oil from the oil cooler maintains full master cylinders by flowing through a brake reservoir. Brake operation can be accomplished with engine inoperative because of the reservoir that keeps master cylinders filled.

ADJUSTMENT

All Models

126. BRAKE PEDAL STOP ADJUSTMENT. To adjust the pedal return stops, remove the right instrument panel side cover. Refer to Fig. 163 and loosen locknuts (5). Remove the stop bolts (2). Lock the pedals together with locking bar (6). One pedal will be in contact with "surface A" of the brake switch mounting bracket. Install the stop bolt and locknut for the pedal that is not in contact with "surface A." Adjust stop bolt until the pedal locking bar moves freely without moving either pedal. Tighten locknut. Install the opposite stop bolt and locknut and adjust bolt until it just contacts its pedal. Tighten the locknut.

127. MASTER CYLINDER PUSH ROD ADJUSTMENT. With the brake pedals against return stop bolts as in paragraph 126, refer to Fig. 163 and loosen clevis locknuts (7). Adjust the two clevises so there is 1.0 mm (0.040 in.) of free travel between the end of the push rods and master cylinder pistons. Tighten the clevis locknuts.

128. BRAKE PEDAL MAXIMUM TRAVEL. To adjust the brake pedal maximum travel, disconnect the brake lines. Refer to Fig. 163 and loosen locknut (9). Turn right maximum travel adjusting bolt (1) counterclockwise until end of bolt is even with mounting bracket face. Apply a pressure of 50 N (11.25 lbs.) to right brake pedal until the master cylinder piston bottoms out. Turn adjusting bolt (1) until bolt just makes contact with the brake pedal. Then, turn adjusting bolt (1) clockwise one complete turn. Tighten the locknut. Repeat the procedure to adjust the left brake pedal. Connect brake lines.

129. BRAKE PEDAL SWITCH ADJUSTMENT. With the brake pedals against the return stop bolts as outlined in paragraph 126, refer to Fig. 163 and disconnect the electrical connector from right brake switch (3). Connect multimeter leads to terminals A and B of brake switch and check for continuity.

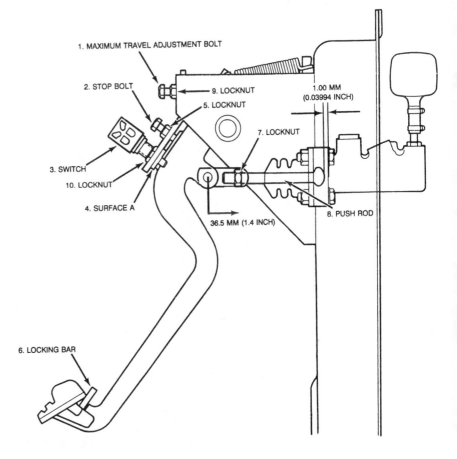

Fig. 163—View showing brake pedal adjusting points. Refer to text for procedures.

Loosen locknut (10) and carefully turn brake switch counterclockwise until there is no continuity. Disconnect multimeter leads. Carefully turn brake switch clockwise ¼ turn to a maximum of 1¼ turns. Tighten locknut and install connector. Repeat procedure for left brake pedal switch.

NOTE: After adjusting the brake pedal switches, terminals on the right switch should face to the right and left switch terminals should face left.

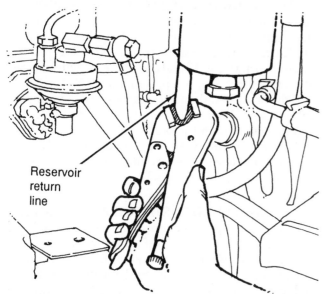

Fig. 164—Use smooth-jawed locking pliers to squeeze off brake return hose.

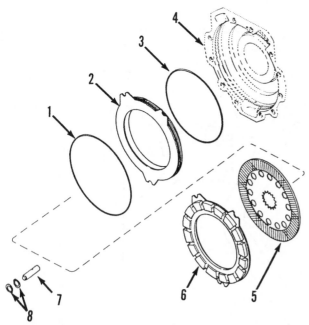

Fig. 165—Exploded view of brake assembly used on all models.

1. "O" ring	5. Friction plate
2. Brake piston	6. Backing plate
3. "O" ring	7. Reaction pin (3)
4. Rear main frame	8. Snap ring

If correct continuity readings cannot be obtained, renew brake switch.

BLEED BRAKES

All Models

130. To bleed air from brake system, open both bleed valves ¼ to ½ turn. Install to each bleed valve a length of 5 mm (³⁄₁₆ in.) ID clear plastic tubing. Place the other ends of tubing into transmission oil filler tube. Start engine and operate at low idle speed. Using smooth-jawed locking pliers, clamp off brake return hose as shown in Fig. 164. With tractor running, brake system will "power bleed" itself. When no air bubbles are seen in the plastic tubes, brake system has been properly bled. Tighten bleed screws and remove plastic tubes. Stop engine and remove locking pliers.

BRAKE ASSEMBLIES

All Models

131. R&R AND OVERHAUL. To remove either brake assembly, first remove final drive as outlined in paragraph 124. Then, remove backing plate (6—Fig. 165), friction plate (5) and withdraw the sun gear shaft. Remove the three reaction pins (7). Remove the brake bleed screw and inject compressed air through the bleeder port to remove brake piston (2). Remove "O" rings (1 and 3) from the piston.

Clean and inspect all parts and renew any showing excessive wear or other damage. Disassemble the three reaction pins as shown in Fig. 166. Free length of spring must be 49.25 mm (1.939 in.). Renew springs if free length is less than the specified amount. Check the snap rings for tightness on the reaction pin. Renew the snap rings if they are loose on pin. Space the eyes of snap rings 180° apart and reassemble reaction pins.

When reinstalling, lubricate new "O" rings (1 and 3—Fig. 165) with clean Hy-Tran Plus oil and install

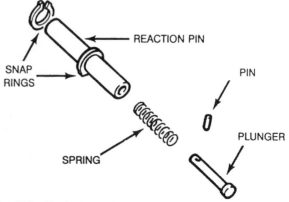

Fig. 166—Exploded view of reaction pin assembly.

on piston (2). Install piston into rear main frame so that lugs are aligned with holes for reaction pins. Install the sun gear shaft. Install the three reaction pins into holes in rear main frame. Refer to Fig. 167 and place a straightedge across the end of the reaction pin plunger and measure distance from the machined surface of the rear main frame to end of plunger. This measurement must be 2.0 mm (0.079 in.). If not, move snap rings on reaction pin to obtain the correct measurement. Repeat the procedure for the other reaction pins. Install friction plate (5—Fig. 165) and brake backing plate (6). Reinstall final drive as outlined in paragraph 124. Refill transmission with Hy-Tran Plus oil to correct level on dipstick. Bleed brakes as outlined in paragraph 130.

BRAKE MASTER CYLINDERS

All Models

132. R&R AND OVERHAUL. To remove brake master cylinders, remove right side instrument panel cover. Raise hood and remove exhaust pipe extension and rear hood cover. Disconnect push rod clevises (23—Fig. 168) from brake pedals. Disconnect brake reservoir supply hose (3) and return hose (2). Disconnect master cylinder supply hoses (5) and remove reservoir (4). Disconnect hydraulic brake tubes and master cylinder compensator tube. Unbolt and remove master cylinder. Remove second master cylinder in the same manner.

To disassemble master cylinder, remove clevis (23), locknut (22) and rubber boot (21). Remove snap ring (20) and push rod (19) with washer. Withdraw piston (17) with seal (18), return spring (16), spring washer (15) and seal (14) from body (13). Remove flow valve adapter (8) with seal ring (9), flow valve (10), seal (11) and ball (12). Remove adapter (6) and screen (7).

Clean and inspect all parts. If cylinder bore or piston is excessively worn or scored, renew complete

master cylinder. A seal kit, consisting of seal ring (9), seal (11), ball (12), seal (14), spring washer (15), seal (18) and snap ring (20), is available for resealing the master cylinder.

Reassemble master cylinder by reversing disassembly procedure. Tighten inlet adapter (6) and flow valve adapter (8) to 41-47 N•m (31-34 ft.-lbs.) torque.

Install master cylinder and tighten retaining bolts to 26-31 N•m (19-23 ft.-lbs.) torque. Refer to paragraph 128 and adjust brake pedal maximum travel before operating brake pedals. Bleed brake system as outlined in paragraph 130.

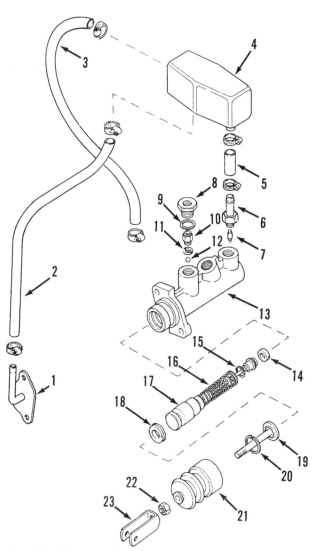

Fig. 168—Exploded view of one master cylinder assembly and related components. Other master cylinder is identical.

1. Flywheel housing cover	9. Seal ring	16. Return spring
2. Brake return hose	10. Flow valve	17. Piston
3. Brake supply hose	11. Seal	18. Piston seal
4. Reservoir	12. Ball	19. Push rod
5. Hose	13. Master cylinder body	20. Snap ring
6. Adapter		21. Rubber boot
7. Screen	14. Valve seal	22. Locknut
8. Adapter	15. Spring washer	23. Clevis

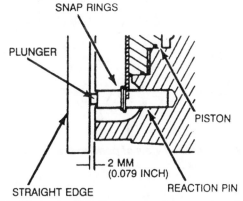

Fig. 167—View showing brake reaction pin adjustment. Refer to text for procedure.

PARKING BRAKE

A mechanically operated, disc-type parking brake, located on front end of bevel drive pinion shaft, is used on all models. When parking brake is engaged, a red warning light will illuminate. An audible warning also sounds if a gear is engaged with parking brake engaged.

ADJUSTMENT

All Models

133. To adjust the parking brake, loosen adjusting nuts (1—Fig. 169). Adjust the length of cable (2) by rotating adjusting nuts until brake lever (3) clicks eight to ten times when brake is applied. Tighten nuts (1) against the bracket.

Check the movement of the switch (4) as parking brake lever (3) is released. After initial contact, the switch plunger should move 10 mm (0.4 in.). If necessary, loosen locknut (5), adjust switch position, then retighten locknut.

R&R AND OVERHAUL

All Models

134. To remove the parking brake assembly, first split tractor between speed transmission and range

transmission as outlined in paragraph 112. On models equipped with front-wheel drive, remove front drive clutch, idler gear and drive gear. On two-wheel-drive models, unbolt and remove the spacer. On all models, remove brake housing (1—Fig. 170) with items (3 through 8). Remove brake hub (9). Separate backing plate (8), brake discs (3 and 7) and actuator assembly (6) from housing (1).

Clean and inspect all parts for excessive wear or other damage. Thickness of new brake discs (3 and 7) is 6.25-6.40 mm (0.246-0.252 in.). Thickness of new back plate (8) is 2.16-2.31 mm (0.085-0.091 in.).

When reinstalling, tighten 12 mm bolts to a torque of 125-150 N·m (92-110 ft.-lbs.) and 16 mm bolts to 310-380 N·m (228-280 ft.-lbs.). The balance of installation is the reverse order of removal procedure.

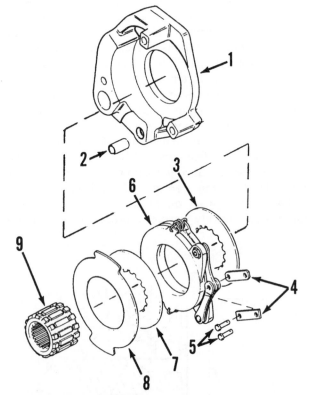

Fig. 170—Exploded view of parking brake components.

1. Brake housing	
2. Dowel	6. Actuator
3. Brake disc	7. Brake disc
4. Links	8. Backing plate
5. Pins	9. Brake hub

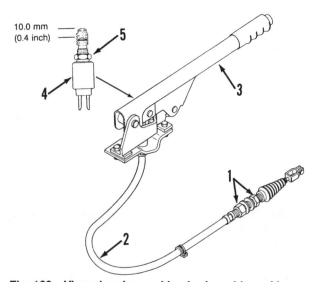

10.0 mm (0.4 inch)

Fig. 169—View showing parking brake cable and lever.

1. Adjusting nuts	
2. Cable	4. Switch
3. Lever	5. Locknut

HYDRAULIC SYSTEM

The hydraulic system provides power for the steering, brakes, pto clutch and brake, Powershift clutches, forward-reverse shuttle clutches, front-wheel drive clutch (if so equipped), differential lock, hydraulic lift, remote valve system, as well as lubrication for various systems. The charge/lubrication pump is located in rear side of range transmission front bearing carrier. The main hydraulic pump is located on right side of speed transmission housing. Refer to paragraph 13 for service information on hydraulic fluid and filter.

Refer to the following paragraphs for hydraulic pumps and system components.

HYDRAULIC PUMPS

All Models

135. PRESSURE AND FLOW TESTS (CHARGE/LUBRICATION PUMP). To test the charge/lubrication pressure, refer to Fig. 171 and remove test port plug from hydraulic oil filter head. Install a 1000 kPa (150 psi) pressure gauge to the test port. Start engine and heat hydraulic oil to a temperature of 50° C (120° F). With engine operating at 2200 rpm, fully cycle inching (clutch) pedal one time. Pressure gauge should read 345-480 kPa (50-70 psi). If not, check for faulty charge pump (low flow) or a faulty relief valve.

To test the charge pump flow, remove the hydraulic oil filter element and by-pass valve. Refer to Fig. 172 and install the flowmeter adapter (CAS-2009-9) as shown. Connect a hydraulic flowmeter (CAS-10280 or equivalent) to inlet and outlet fittings. With engine operating at 2200 rpm and oil temperature at 50° C (120° F), flow reading should be 102-121 L/m (27-32

U.S. gpm). If not, check for a faulty charge pump or a clogged suction screen.

136. PRESSURE AND FLOW TESTS (MAIN HYDRAULIC PUMP). To check the main pump low standby pressure, refer to Fig. 173 and remove test port plug from bottom of pump compensator. Install

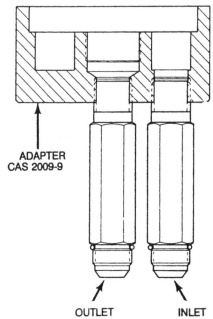

Fig. 172—Flowmeter adapter (CAS-2009-9) installed in place of hydraulic oil filter element to test charge pump flow.

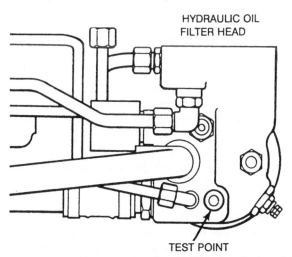

Fig. 171—Top view of hydraulic oil filter head showing charge and lubrication pump pressure test port.

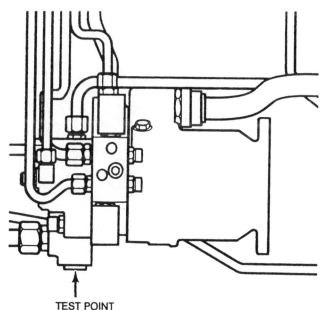

Fig. 173—Install pressure gauge at test point shown when checking low standby pressure or pump maximum pressure.

a 20,000 kPa (3000 psi) pressure gauge in test port. With engine operating at 2200 rpm, hydraulic oil temperature at 50° C (120° F), and hitch and remote valves in neutral, turn steering wheel to full lock and return to neutral. The low standby pressure must be 2760-4140 kPa (400-600 psi). If pressure is too high, check for blocked orifice in steering hand pump or a blocked bleed orifice in the remote valve block. If pressure is too low, flow compensator is not adjusted correctly or internal orifice in main pump is blocked.

To test the main pump regulated pressure, remove plug or disconnect power shift supply line as shown

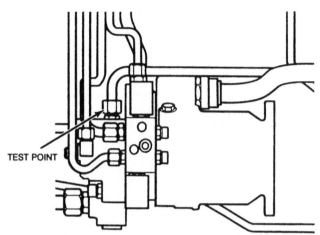

Fig. 174—Remove plug or disconnect power shift supply line and install pressure gauge in test port to test regulated pressure.

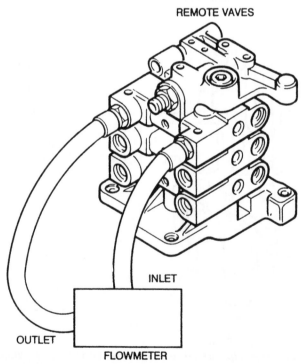

Fig. 175—Connect a flowmeter to a remote valve as shown to check main pump flow.

in Fig. 174. Then, install a 4000 kPa (600 psi) pressure gauge in the test port. With engine operating at 1500 rpm and hydraulic oil temperature at 50° C (120° F), fully cycle inching pedal one time. The pressure reading must be 1800-2070 kPa (260-300 psi). If pressure is not correct, compensator is faulty and must be renewed.

To test main pump maximum pressure, refer to Fig. 173 and remove test port plug from bottom of pump compensator. Install a 20,000 kPa (3000 psi) pressure gauge in test port. With engine operating at 1500 rpm and hydraulic oil temperature at 50° C (120° F), hold remote valve control lever fully on for maximum pressure. Maximum pressure reading should be 18,600-19,300 kPa (2700-2800 psi). If pressure reading is not correct, high-pressure relief valve or flow compensator requires adjustment or main pump is worn.

To test the main pump flow, refer to Fig. 175 and connect a flowmeter to a remote valve as shown. Heat hydraulic oil to a temperature of 50° C (120° F). Turn flow control knobs to maximum position. With engine operating at 2200 rpm, hold remote valve lever fully on for maximum flow. Adjust flowmeter to 3450 kPa (500 psi). Flow should be 60 L/m (16 U.S. gpm). Then, adjust flowmeter to 13,790 kPa (2000 psi). Flow should be 57 L/m (15 U.S. gpm). If flow is less than specified, check for worn main pump, bad flow control valve or faulty steering priority valve in compensator block.

137. R&R AND OVERHAUL CHARGE/LUBRI-CATION PUMP. To remove the charge/lubrication pump, first split tractor between speed transmission and range transmission and remove range transmission front bearing carrier as outlined in paragraph 112. Place bearing carrier housing on a clean bench.

To disassemble the charge/lubrication pump, remove snap ring (20—Fig. 176), thrust washers (18) and thrust bearing (19). Remove snap ring (3), thrust washers (4) and thrust bearing (5). Unbolt and remove scavenger pump housing (7). Remove scavenger pump gerotor set (10), spacer plate (11) and shims (12). Remove drive shaft (15), charge/lubrication pump gerotor set (13) and wear plate (14). Using a blind hole puller, such as (CAS-10355), remove needle bearing (16) from pump side of bearing carrier housing. Press needle bearing (9) out gerotor side of housing (7).

Clean and inspect all parts and renew any showing excessive wear or other damage. Light scratches on wear plate (14) and spacer plate (11) can be polished out with fine emery cloth.

When reassembling, lubricate all parts with clean Hy-Tran Plus oil. Install needle bearing (16) into bearing carrier housing (17) until bearing cage is flush with bottom of chamfer. Install wear plate (14) so that tab on plate locates in hole in bearing carrier housing. Install charge/lubrication pump gerotor set (13). Place a straightedge across gerotor set as shown in Fig. 177, then, using a feeler gauge, measure gap between bearing carrier housing and straightedge. Add 0.038-0.076 mm (0.0015-0.0030 in.) to the gap measurement to determine correct shim pack thickness to be installed. Shims are available in two thicknesses: RED, 0.05 mm (0.002 in.) and BLUE, 0.13 mm (0.005 in.). Install drive shaft (15—Fig. 176), then install the selected shim pack (12) and spacer plate (11). Place the scavenger pump gerotor set (10) in the housing (7). Place a straightedge across the housing and, using a feeler gauge, measure the clearance between the gerotor set and the straightedge. The clearance should be 0.087-0.163 mm (0.003-0.006 in.). If clearance exceeds the specified amount, renew gerotor set (10) and/or housing (7). Press needle bearing (9) into bore in housing (7) until seated against shoulder. Install scavenger pump gerotor set (10) over the drive shaft. Install

housing (7) and tighten bolts (6) to a torque of 134-151 N·m (100-110 ft.-lbs.). Install new "O" ring in inlet port of housing (7). Install thrust washers (4) with thrust bearing (5) between washers, then install snap ring (3). Install thrust washers (18) with thrust bearing (19) between the washers, then install snap ring (20). Refer to Fig. 178 for cross-sectional view of charge/lubrication pump assembly.

Refer to paragraph 112 and reinstall range transmission front bearing carrier and reassemble tractor by reversing the disassembly procedure.

138. R&R AND OVERHAUL MAIN PUMP, COMPENSATOR VALVE AND DRAFT CONTROL VALVE. To remove the main hydraulic pump, disconnect battery cables and drain transmission oil. Support tractor and remove right rear wheel. Unbolt and remove protective shield and plate from right side of transmission. Identify, tag and disconnect electrical connectors from the pressure switch and draft control solenoids. Remove tubes from hitch input control valve and tube from draft control valve. On models so equipped, remove tube to power shift manifold. On all models, remove tube to remote valves and the tube to position sensing valve from draft control valve. Remove hitch lift tube from draft control valve and the tubes to steering hand pump. Remove pressure compensator tube to remote valve manifold and loosen hose clamps on filter to pump supply tube. Unbolt and remove main pump assembly from tractor. Unbolt and remove pump supply tube from pump.

Separate compensator valve (3—Fig. 179) and draft control valve (2) from main pump (1) as follows: The main pressure relief valve (5) and steering pressure relief valve (6) are identical except for adjustment and location. Mark location of relief valves (5 and 6), then remove assemblies. Remove the four Allen screws (4), then separate the components.

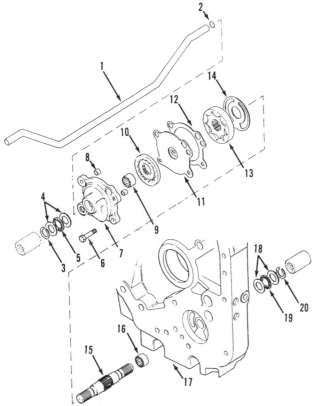

Fig. 176—Exploded view of charge/lubrication pump and related components.

1. Inlet tube	11. Spacer plate
2. "O" ring	12. Shim
3. Snap ring	13. Charge/lubrication
4. Thrust washers	pump gerotor set
5. Thrust bearing	14. Wear plate
6. Bolt	15. Drive shaft
7. Scavenger pump	16. Needle bearing
housing	17. Bearing carrier
8. Dowel (2)	housing
9. Needle bearing	18. Thrust washers
10. Scavenger pump	19. Thrust bearing
gerotor set	20. Snap ring

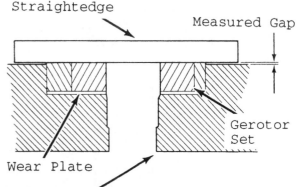

Fig. 177—Measure gap as shown when determining shim pack thickness to be installed on charge/lubrication pump.

Refer to the appropriate following paragraphs for disassembly and overhaul procedures for compensator valve, draft control valve and main hydraulic pump.

139. COMPENSATOR VALVE. To disassemble the compensator valve, remove the pressure switch from the tee fitting (34—Fig. 180). Remove the tee fitting and adapter (32). Remove remote valve pilot adapter (25), check valve seat and ball (27 and 28). Remove the manifold supply adapter (38) and screen (40). Remove the steering pilot tube adapter (42) and the remote valve supply tube adapter (36). Remove compensator spool adjuster (47 through 51), springs (45 and 46) and spring seat (44). Remove plug (23) and compensator spool (21). Remove cap (52) and spring (54), then remove plug (20) and steering priority spool (18). Remove plug (55), spring (57) and regulator valve spool (58). Remove plug (17), orifice (16) and plug (15). Remove Allen head plugs from clean-out holes in compensator body (12). To disassemble the main pressure relief valve (M) or the steering pressure relief valve (S), loosen locknut (1). Then, using an Allen wrench, remove adjustment screw (2). Remove spring (3) and spring seat (4). Remove valve stem (6) with snap ring (5).

Clean and inspect all parts for excessive wear or other damage. Use compressed air to blow out all passages in body (12). Renew compensator assembly if there is excessive wear or damage.

When reassembling, use Loctite 242 on all clean-out plugs. Use all new "O" rings and reassemble in reverse order of disassembly procedure.

After compensator valve assembly is reinstalled on tractor, adjust main pressure relief valve (5—Fig.

179) and steering relief valve (6) as outlined in paragraph 15.

140. DRAFT CONTROL VALVE. Remove the draft control valve as outlined in paragraph 138. To disassemble the draft control valve, refer to Fig. 181 and place identification marks on solenoid valves for aid in reassembly. Remove nuts (15 and 32), "O" rings (16 and 31), coils (17 and 30) and "O" rings (18 and 29). Unscrew and remove the raise solenoid valve (19) and lower solenoid valve (28) with "O" rings (20 and 27). Remove the four Allen screws (22) and lift off port housing (21). Remove spring seat (13), centering spring (12), spool (11), collar (10), "O" ring (9) and centering spring (40). Remove spring seat (33), spring (34) and "O" ring (35). Remove spool (8) and "O" ring (7). Remove the six Allen screws (42) and remove side housing (41). Remove spring seat (2), outer spring (4), inner spring (3), drop check valve (5) and pin (6). Remove plug (39) with "O" ring (38) and flow compensator spool (37). Then, using a 2.5 mm Allen wrench, remove plug, check ball and spring from inside of compensator spool. Loosen locknut (24) and, using a

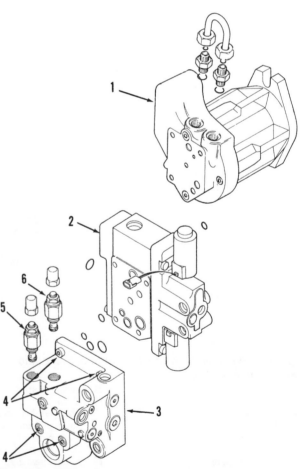

Fig. 179—View showing compensator valve and draft control valve removed from main hydraulic pump.

1. Main pump	4. Allen screws
2. Draft control valve	5. Main pressure relief valve
3. Compensator valve	6. Steering pressure relief valve

Fig. 178—Cross-sectional view of charge/lubrication pump assembly.

Bearing Carrier Housing
Spacer Plate
Scavenge Pump Gerotor Set
Scavenge Pump Housing
Bearing Snap Ring
Thrust Washers
Snap Ring
Drive Shaft
Bearing
Bearing
Thrust Washers O-Ring
Bearing
Scavenge Oil Inlet Tube
Shim
Charge/Lubrication Pump Gerotor Set

3 mm Allen wrench, remove spool centering screw (23). Then, using a 4 mm Allen wrench, remove drop rate adjusting screw (14). Remove line adapters and clean-out plugs.

Clean all parts in cleaning solvent and dry with compressed air. Renew the complete valve assembly if any wear or other damage is found.

When reassembling, renew all "O" rings and reassemble in reverse order of disassembly procedure, keeping the following points in mind. When installing hitch drop rate screw (14), end of adjusting screw must be 2-3 mm (0.078-0.118 in.) below machined

surface of port housing (21). When installing clean-out plugs, apply Loctite 242 to the plug threads. Apply Loctite 242 to threads of check ball plug in flow compensator spool (37). Tighten Allen screws (42) to a torque of 32 N·m (23 ft.-lbs.) and tighten Allen screws (22) to 62 N·m (45 ft.-lbs.). Tighten solenoid valves (19 and 28) to a torque of 15-18 N·m (11-18 ft.-lbs.). Refer to paragraph 157 for draft control valve adjustment.

141. MAIN HYDRAULIC PUMP. Remove main hydraulic pump as outlined in paragraph 138 and

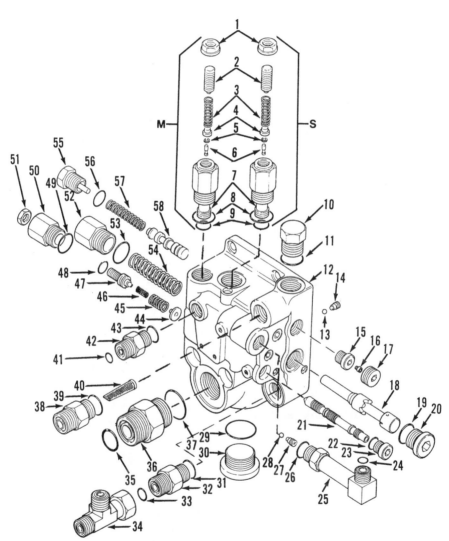

Fig. 180—Exploded view of compensator valve assembly.

M. Main pressure relief valve	9. "O" ring		
	10. Plug	20. Plug	31. "O" ring
S. Steering pressure relief valve	11. "O" ring	21. Compensator spool	32. Adapter
	12. Compensator body	22. "O" ring	33. "O" ring
1. Locknut	13. Check valve ball	23. Plug	34. Tee fitting
2. Adjusting screw	14. Check valve seat	24. "O" ring	35. "O" ring
3. Spring	15. Plug	25. Adapter	36. Adapter
4. Spring seat	16. Orifice	26. "O" ring	37. "O" ring
5. Snap ring	17. Plug	27. Check valve seat	38. Adapter
6. Valve stem	18. Steering priority spool	28. Check valve ball	39. "O" ring
7. Relief valve body	19. "O" ring	29. "O" ring	40. Screen
8. "O" ring		30. Plug	41. "O" ring

42. Adapter		
43. "O" ring		51. Locknut
44. Spring seat		52. Cap
45. Spring (outer)		53. "O" ring
46. Spring (inner)		54. Spring
47. Adjustment screw		55. Plug
48. "O" ring		56. "O" ring
49. "O" ring		57. Spring
50. Compensator valve adjuster body		58. Regulator valve spool

separate from draft control valve and compensator valve. See Fig. 179. Remove the "U" tube and adapters from pump end cover. Place match marks across end cover (1—Fig. 182) and pump body (8) for aid in reassembly. Remove end cover retaining Allen screws (30) and carefully remove the end cover. Note the position of the orifice in the orifice plate (17) and remove the plate. Remove spring (26), plunger (25) and sleeve (15). Remove bearing cone (18) and spacer (19) from end of pump shaft (6). Remove cylinder block (14) and pistons (11) as an assembly. Grasp the

piston shoeplate (12) and remove pistons (11) from cylinder block (14). Remove dome washer (13) and the three pins (24). Place match marks on swashplate (5) and pump body for aid in correct installation. Remove swashplate assembly from pump body. Remove bearing cups (4) from swashplate, then remove insert (3). Withdraw pump shaft (6) and remove snap ring (10) and oil seal (9). Place cylinder block assembly in a press to compress spring (22) about 1 mm (0.040 in.). Remove snap ring (20) from its groove in cylinder block Release pressure, then remove snap ring (20), washer (21), spring (22) and washer (23).

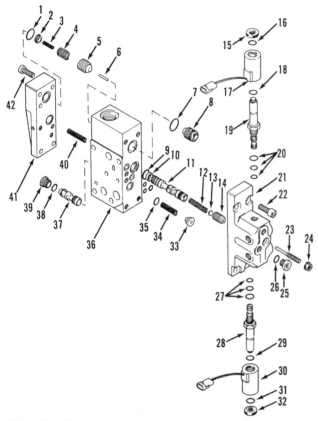

Fig. 181—Exploded view of hitch draft control valve assembly.

1.	"O" ring	22.	Allen screw
2.	Spring seat	23.	Spool centering screw
3.	Spring (inner)	24.	Locknut
4.	Spring (outer)	25.	Plug
5.	Drop check valve	26.	"O" ring
6.	Pin	27.	"O" rings
7.	"O" ring	28.	Solenoid (lower)
8.	Spool	29.	"O" ring
9.	"O" ring	30.	Coil
10.	Collar	31.	"O" ring
11.	Main spool	32.	Nut
12.	Centering spring	33.	Spring seat
13.	Spring seat	34.	Spring
14.	Drop rate adjusting	35.	"O" ring
	screw	36.	Valve body
15.	Nut	37.	Flow compensator
16.	"O" ring		spool
17.	Coil	38.	"O" ring
18.	"O" ring	39.	Plug
19.	Solenoid (raise)	40.	Centering spring
20.	"O" rings	41.	Side housing
21.	Port housing	42.	Allen screw

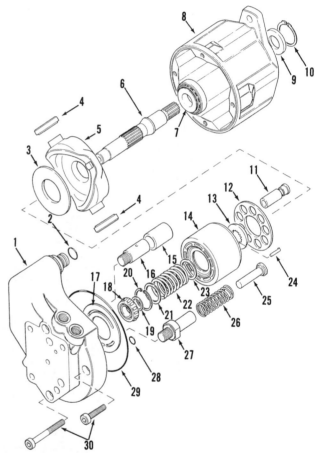

Fig. 182—Exploded view of main hydraulic pump assembly used on all models.

1.	End cover	16.	Pin
2.	"O" ring	17.	Orifice plate
3.	Insert	18.	Bearing cone
4.	Bearing cup & cone	19.	Spacer
5.	Swashplate	20.	Snap ring
6.	Pump shaft	21.	Washer
7.	Bearing cup & cone	22.	Spring
8.	Pump body	23.	Washer
9.	Oil seal	24.	Pin (3)
10.	Snap ring	25.	Plunger
11.	Piston (9)	26.	Spring
12.	Shoe plate	27.	Pin
13.	Dome washer	28.	"O" ring
14.	Cylinder block	29.	"O" ring
15.	Sleeve	30.	Allen screws

Clean all parts of the hydraulic pump in cleaning solvent and dry with compressed air. Renew complete pump assembly if any parts are worn or damaged.

When reassembling, use all new "O" rings and oil seal. Reassemble in reverse order of disassembly procedure, keeping the following points in mind. Lubricate bearings (4, 7 and 18), sleeve (15), plunger (25), the three pins (24) and dome washer (13) with petroleum jelly. Lubricate insert (3), orifice plate (17) and pistons (11) with clean Hy-Tran Plus oil. Tighten Allen screws (30) to a torque of 62 N·m (45 ft.-lbs.).

Reinstall draft control valve (2—Fig. 179) and compensator valve (3) on main pump (1) and tighten Allen screws (4) to a torque of 50 N·m (37 ft.-lbs.). Install "U" tube adapters on pump, then pour 0.65 liter (0.7 U.S. qt.) of Hy-Tran Plus oil through adapters into the pump. Install "U" tube. Use new mounting gasket and install the assembly. Tighten pump retaining bolts to a torque of 134-151 N·m (100-110 ft.-lbs.). Use new "O" rings on tube fittings and complete installation in reverse order of removal. Fill transmission to correct level on dipstick with Hy-Tran Plus oil. Adjust main pressure relief valve (5—Fig. 179) and steering pressure relief valve (6) as outlined in paragraph 15. Adjust draft control valve as outlined in paragraph 157. Install rear wheel and tighten mounting nuts on flange axle to 434-475 N·m (320-350 ft.-lbs.) or mounting bolts on rowcrop axle to 300 N·m (220 ft.-lbs.).

PTO FRONT DRIVE SHAFT AND HYDRAULIC PUMP DRIVE GEAR

All Models

142. R&R AND OVERHAUL. To remove the pto front drive shaft and main hydraulic pump drive gears, first remove main pump assembly as outlined in paragraph 138. Unbolt and remove hydraulic housing and filter. Split tractor and remove speed transmission input shaft and drop shaft as outlined in paragraph 108 (Powershift) or paragraph 110 (synchromesh). Then, drive plug (22—Fig. 183) forward out of transmission housing. Remove snap ring (14) and drive pump coupler shaft (17), bearing cone (16) and bearing cup (15) rearward out of housing. Remove pump drive gear (18), bearing cone (19) and snap ring (21), then drive out bearing cup (20). Remove bolt (13), washer (12), shim (11) and pto drive gear (10). Remove snap ring (3) from its groove in shaft (2). Install a slide hammer and adapter in front end of shaft and pull shaft (2) and bearing cone (9) forward out of housing. Remove coupling (1) and bearing cone (4). Remove bearing cup (8), snap rings (6 and 7) and bearing cup (5).

Clean and inspect all parts for excessive wear or other damage and renew as necessary.

When reinstalling, install snap ring (3) into its groove in shaft (2). Install bearing cone (4) until seated against snap ring (3). Install coupling (1) and shaft (2) into the transmission housing. Install bearing cup (5) and snap ring (6). Pull shaft (2) forward to seat bearing cup (5) against snap ring (6). Install snap ring (7), then install bearing cup (8) until seated against the snap ring. Install bearing cone (9), gear (10) and washer (12). Do not install shims (11) at this time. Install and tighten bolt (13) finger tight. While rotating shaft (2), tighten bolt (13) until bearings (4 and 9) have an end play of 0.076 mm (0.003 in.), measured with a dial indicator against washer (12). Measure gap between end of shaft (2) and washer (12) with a feeler gauge through slot in gear (10). This measurement will be the thickness of shim pack (11) to be installed. Shims are available in thicknesses of 0.076, 0.305 and 1.017 mm (0.003, 0.012 and 0.040 in.). Remove bolt and washer, then install shim pack, washer and bolt. Tighten bolt to a torque of 125-150 N·m (93-112 ft.-lbs.). Using a dial indicator against washer (12), recheck bearing end play. End play must be 0.025-0.125 mm (0.001-0.005 in.). If not, recheck shimming procedure.

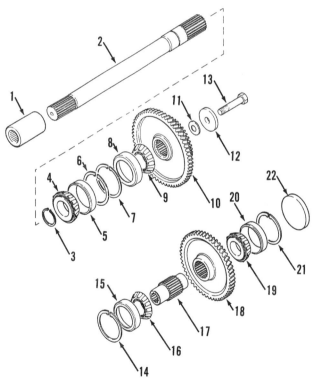

Fig. 183—Exploded view of pto front drive shaft and main hydraulic pump drive gears.

1. Coupling	9. Bearing cone	17. Pump coupler
2. Pto drive shaft	10. Drive gear (pto)	shaft
3. Snap ring	11. Shim	18. Drive gear
4. Bearing cone	12. Washer	(pump)
5. Bearing cup	13. Bolt	19. Bearing cone
6. Snap ring	14. Snap ring	20. Bearing cup
7. Snap ring	15. Bearing cup	21. Snap ring
8. Bearing cup	16. Bearing cone	22. Plug

Press bearing cone (19) onto pump coupler shaft (17) until seated against splines. Install gear (18) on the coupler. Install a 2.0 mm (0.079 in.) snap ring (14), bearing cup (15) and bearing cone (16). Drive coupler (17) rearward into bearing cone (16) until the bearing cone is seated against coupler splines. Press bearing cup (20) against bearing cone (19) while rotating coupler (17), until all end play is eliminated. Install thickest snap ring (21) possible against bearing cup (20). Snap rings are available in thicknesses of 1.5, 1.7 and 2.0 mm (0.059, 0.067 and 0.079 in.). With end play of bearings eliminated, use a feeler gauge and measure gap between snap ring (21) and bearing cup (20). This measurement must be between 0.025 and 0.127 mm (0.001 and 0.005 in.). If measurement is incorrect, remove snap ring (21) and install a snap ring of correct thickness to provide the required measurement. Install plug (22) until lip of plug is recessed 3 mm (0.125 in.) below face of bore.

The balance of reassembly is the reverse order of the disassembly procedure.

HYDRAULIC LIFT SYSTEM

REAR HITCH OPERATION

All Models

143. All tractors are equipped with a hydraulic hitch system that permits the position and load control to be operated by a single setting lever (Fig. 184). The mode selector is used to select either load or position control. In position control, the operator uses the setting lever to select the working position of the implement. In load control, the operator uses the setting lever to adjust the load on the tractor of ground engaging implements. The hydraulic system will automatically adjust the hitch position to maintain a constant implement load on the tractor as the implement travels through varying soil conditions and terrain. If the implement travels through heavy soil, the increased load is sensed at the lower links and the hitch will raise the implement in small increments to maintain a constant load on the tractor. As the implement travels through lighter soil conditions, the lower links will sense the load decrease and lower the implement in small increments while maintaining a constant load on the tractor. In load control, the hitch can move through the full movement if necessary to maintain constant load on the tractor.

144. MODE SELECTOR. The mode selector (Fig. 184) has three marked positions. The three functions are position, mix and load (draft). To operate the mode selector, rotate the outside of the knob.

When the mode selector is in position control, the setting lever fully controls the hitch position without any load control. The position selection can be used when implement depth is critical or on above ground equipment.

When mode selector is in mix control, the setting lever controls the hitch position and the mode selector

controls load sensitivity within a 60° range. With the dial set close to position, there is little load sensitivity and the implement will closely maintain the set depth. With the dial set close to load, there is more load sensitivity and less position control, so the implement will vary more from the set depth. To use the mix selection, start with mode selector in position. Use the setting lever to adjust the depth desired. Then, turn the mode selector into mix position, turning selector counterclockwise until the desired amount of load sensitivity is reached.

With the mode selector in load (draft) control, the setting lever fully controls the load sensitivity, without any lower set depth. Load selection can be used where soil conditions are consistent or on implements with gauge wheels. If draft load decreases, implement depth increases.

145. ENABLE INDICATOR LAMP. When the tractor is started, the enable indicator lamp (Fig. 185) will flash, which indicates that the system is not enabled and the hitch will not operate. Enable the system as follows: When the engine is first started and hitch command switch is in full forward or full rearward detented positions, it must be moved to the center (neutral) position, then back to working or full raise (transport) position as desired. If the lever is in the center (neutral) position, move it to the working or full raise position as desired. The hitch must be enabled each time the engine is started to keep the hitch in the desired position.

146. HITCH COMMAND SWITCH. When the five-position hitch command switch (Fig. 186) is in the working position, the hitch will react to the set-

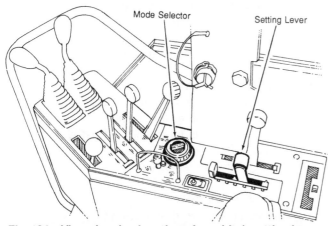

Fig. 184—View showing location of rear hitch setting lever and mode selector on right console.

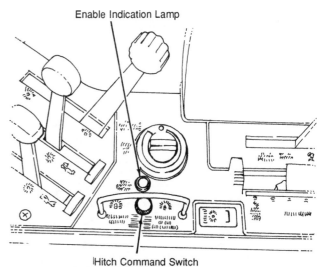

Fig. 185—View showing location of rear hitch enable light and hitch command switch.

ting lever and mode selector positions. This position should be used to place the implement in its working position. In lower position, the hitch will lower regardless of setting lever and mode control positions. In neutral position, the hitch is not operable. Move command lever to working or full raise position. When in raise position, the hitch will raise regardless of setting lever and mode selector positions. This can be used for coupling up to implements. In full raise (transport) position, the hitch will move to fully raised position. This position should be used to raise the implement at the headland or for transport of a mounted implement.

> CAUTION: The purpose of the enable system is to avoid accidental movement of the hitch when starting the engine. Before moving the hitch command switch, make certain there is no person in the area near the rear 3-point hitch or implement.

147. RATE OF DROP CONTROL. This unit controls the speed at which the hitch lowers and is operated by turning the center of the mode selector knob. Turning the knob clockwise increases the low-

ering speed, and turning the knob counterclockwise decreases the lowering speed. This control only operates when hitch command switch is in the working (fully forward) position.

148. SETTING LEVER. The setting lever has a scale of "0" to "10," "0" being raise or light and "10" being lower or heavy. When working in Position Control, the setting lever is used to adjust the position of the implement. In Position Control, to lift an implement, move the lever rearward. To lower an implement, move the lever forward. When working in Load Control, the setting lever is used to adjust the load on the tractor when using ground engaging implements. In Load Control, to decrease the load, move the lever rearward. To increase the load, move the lever forward.

149. MECHANICAL STOP. Mechanical stops (Fig. 187) are used to limit lift or lower positions of setting lever. Adjust the stops by turning the thumb wheels. The rear thumb wheel adjusts the front stop and the front thumb wheel adjusts the rear stop.

TROUBLESHOOTING THE HITCH SYSTEM

All Models

150. The following are symptoms that may occur during operation of the rear hydraulic hitch system and some possible causes of the symptoms. Using this information in conjunction with the Adjustment, R&R and Overhaul information, should help in servicing the hydraulic hitch system.

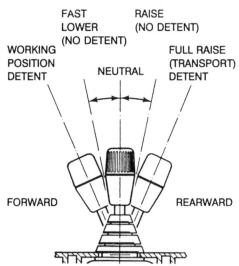

Fig. 186—View showing the five positions of the hitch command switch.

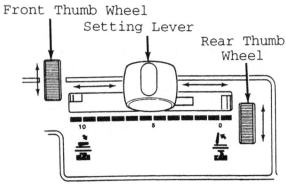

Fig. 187—Mechanical stops are used to limit travel of the setting lever. Rotate the thumb wheels to adjust stops.

1. **Hitch does not respond to setting lever when in position mode.**
   A. Faulty working valve solenoids.
   B. Low regulated pressure.
   C. Main draft spool binding.
   D. Setting lever linkage damaged.
   E. Position sensor orifice blocked.
   F. Position sensor linkage damaged.
   G. Faulty or plugged draft sensors.

2. **Hitch draft operates too slowly.**
   A. Low main pump pressure and/or flow.
   B. Low regulated pressure or flow.
   C. Main draft spool incorrectly adjusted.
   D. Faulty hitch cylinder seals.
   E. Faulty hitch relief valve.

3. **Hitch will not raise.**
   A. Incorrect low standby pressure.
   B. High-pressure relief valve stuck open.
   C. Faulty flow compensator.

D. Input valve incorrectly adjusted.

E. Faulty main pump.

F. Faulty raise solenoid.

G. Faulty hitch cylinder seals.

4. Hitch will not lower.

A. Faulty lower solenoid.

B. Drop rate control valve incorrectly adjusted.

C. Setting lever linkage out of adjustment.

D. Main draft spool incorrectly adjusted or binding.

E. Drop poppet sticking.

F. Flow compensator incorrectly adjusted.

5. Hitch raises too slowly.

A. Low main pump pressure.

B. Input control valve maladjusted.

C. Faulty hitch cylinder seals.

D. Faulty drop rate control valve.

E. Hitch relief valve faulty.

F. Flow compensator incorrectly adjusted.

G. Faulty position control valve.

6. Hitch drops too fast.

A. Input control valve maladjusted.

B. Drop rate control valve linkage incorrectly adjusted.

C. Main draft spool centering screw incorrectly adjusted.

D. Faulty drop rate control valve.

7. Drop rate too slow.

A. Drop rate control linkage maladjusted.

B. Mode selector linkage maladjusted.

C. Main draft spool incorrectly adjusted or sticking.

D. Flow compensator incorrectly adjusted.

E. Faulty drop rate control valve.

8. Hitch leaks down.

A. Faulty drop check valve.

B. Faulty hitch relief valve.

C. Faulty hitch cylinder seals or cracked cylinder.

ADJUSTMENTS

All Models

151. If the hitch fails to operate correctly and must be adjusted, the following adjustment procedure has a logical progression and it should be followed in sequence to achieve the correct results. Before performing the following adjustments, operate tractor until hydraulic oil temperature is 50° C (120° F). Check to see that hydraulic pump pressures are correct and adjust if necessary as in paragraph 136.

152. MAIN CONTROL VALVE SPOOL CENTERING ADJUSTMENT. Install an implement or weight onto hitch with a minimum weight of 1400 kg

(3100 lbs.). Start engine and raise the hitch to the half-way position. Move the hitch command switch to the vertical (neutral) position. Check for any movement of the hitch. If the hitch raises or lowers, the main control valve spool requires centering.

Lower the implement or weight to the ground with the command switch and stop the engine. Loosen the spool centering screw locknut (24—Fig. 188). Start engine and raise the hitch to the half-way position. Use a 3 mm Allen wrench to turn spool centering screw (23) clockwise until the hitch just starts to raise. Make a note of the position of the screw. From

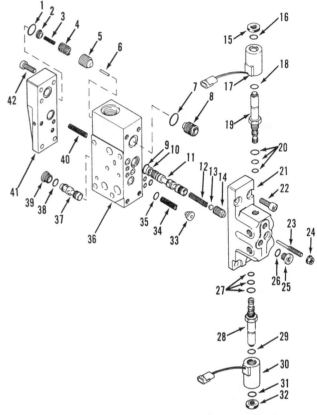

Fig. 188—Exploded view of hitch draft control valve.

1. "O" ring	22. Allen screw
2. Spring seat	23. Spool centering screw
3. Spring (inner)	24. Locknut
4. Spring (outer)	25. Plug
5. Drop check valve	26. "O" ring
6. Pin	27. "O" rings
7. "O" ring	28. Solenoid (lower)
8. Spool	29. "O" ring
9. "O" ring	30. Coil
10. Collar	31. "O" ring
11. Main spool	32. Nut
12. Centering spring	33. Spring seat
13. Spring seat	34. Spring
14. Drop rate	35. "O" ring
adjusting screw	36. Valve body
15. Nut	37. Flow compensator
16. "O" ring	spool
17. Coil	38. "O" ring
18. "O" ring	39. Plug
19. Solenoid (raise)	40. Centering spring
20. "O" rings	41. Side housing
21. Port housing	42. Allen screw

this point, count the number of turns while turning the centering screw counterclockwise until the hitch just starts to lower. Divide the number of turns by two and turn the screw clockwise by that amount. Hold the screw in position and tighten the locknut (24). Lower the hitch, stop the engine and remove the weight or implement.

153. HITCH MAXIMUM HEIGHT ADJUSTMENT. Manually lift the rockshaft arms until the internal stop in the pto housing is contacted. Scribe a reference mark across the hub of the rockshaft arm and the pto housing. Lower the rockshaft. Install an implement or weight onto the hitch with a minimum weight of 1400 kg (3100 lb.). Start the engine and move hitch command switch fully rearward to the full raise detented position. When hitch is fully raised, check the reference marks on the rockshaft arm and the pto housing. The two marks should be 6 mm (0.250 in.) from each other.

If not, lower the implement or weight to the ground. Stop engine and adjust the position sensor check

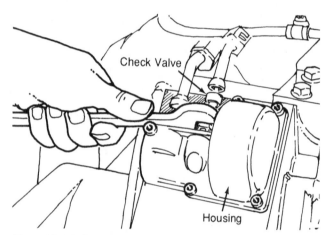

Fig. 189—Adjust the position sensor check valve clockwise to decrease maximum hitch height or counterclockwise to increase maximum hitch height.

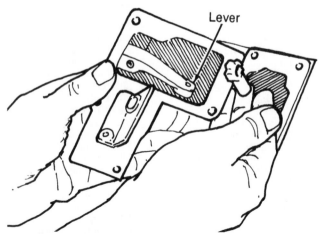

Fig. 190—View showing position sensing connecting rod removed from lever in sensor housing.

valve. This valve is a mechanically operated check valve that limits the rockshaft to within 6° of the internal mechanical stop. Disconnect the hydraulic tubes from the position sensor housing. Remove the left tube adapter. Loosen the locknut on the position sensor check valve. Turn the check valve body (Fig. 189) clockwise a small amount to decrease maximum hitch height or counterclockwise to increase maximum hitch height, then tighten the locknut. Using new "O" rings, install adapter and connect hydraulic tubes. Recheck maximum hitch height and readjust the position sensor check valve as required to provide a distance of 6 mm (0.250 in.) between the two reference marks.

> **NOTE: If hitch maximum height cannot be reached by adjusting check valve, it will be necessary to adjust length of the position sensing connecting rod as outlined in the following paragraph.**

154. POSITION SENSING CONNECTING ROD ADJUSTMENT. If correct results cannot be obtained by adjusting the position sensor check valve as in paragraph 153, adjust the position sensing connecting rod as follows:

With the implement or weight lowered to the ground and engine stopped, disconnect the two hydraulic tubes from the position sensor housing (Fig. 189). Remove the position sensor housing retaining screws. Remove the sensor housing and disconnect the sensing connecting rod from the lever in the housing (Fig. 190). Then, using a prybar through the opening as shown in Fig. 191, remove connecting rod from rockshaft internal lift arm. Adjust the length of the connecting rod so the distance between the two locknuts is 159 mm (6.25 in.). See Fig. 192. Be sure that both end sockets are facing the same direction.

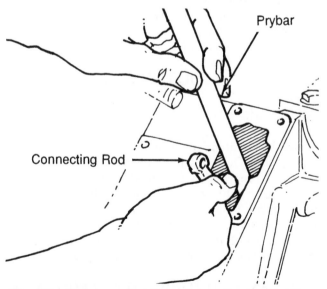

Fig. 191—Use a pry bar through opening as shown to remove position sensing connecting rod from rockshaft internal lift arm.

Reinstall connecting rod. Apply Loctite 515 sealant onto the housing, then install position sensor housing. Tighten retaining screws to a torque of 11-13 N·m (8-10 ft.-lbs.). Reconnect the hydraulic tubes.

NOTE: After completing position sensing connecting rod adjustment, repeat the position sensor check valve adjustment outlined in paragraph 153.

155. INPUT VALVE SETTING LEVER LINKAGE ADJUSTMENT. The input valve is bolted under the cab floor in front of the right rear cab mount. To adjust the setting lever linkage, remove pin and disconnect lower linkage clevis from input valve arm. Remove the mounting bolt from the left side of input valve bracket. Bolt the special lever setting tool (CAS-2032) to the valve bracket as shown in Fig. 193. When the special tool is bolted to the valve bracket, the valve lever is automatically held in the correct setting position. Refer to Fig. 194 and rotate thumb wheels to move lever stops rearward to lock setting lever in "0" position. Loosen the locknut on setting lever linkage and adjust the clevis on the linkage until the clevis pin can be freely inserted through clevis and valve lever. Install pin and secure with clip, then tighten the locknut. Unbolt special setting tool and remove from input valve. Install and tighten valve bracket bolt.

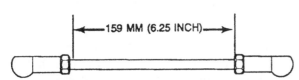

Fig. 192—*Adjust length of connecting rod to dimension shown as measured between the locknuts. Both end sockets must face the same direction.*

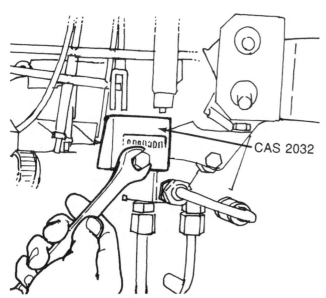

Fig. 193—*Special setting lever linkage adjustment tool (CAS-2032) installed on input valve mounting bracket.*

156. POSITION SENSING ORIFICE ADJUSTMENT. Start engine and set the hitch rockshaft arms to the fully lowered position. Move mode selector clockwise to position mode. Move hitch command switch fully forward to working position detent. Move the setting lever fully forward to position "10."

With the engine running, rockshaft arms should remain in lowered position. Move setting lever to position "9" and rockshaft arms should start to raise. If arms remain stationary, position sensing orifice screw requires adjustment. Loosen locknut on orifice adjusting screw. Refer to Fig. 195 and install a 3 mm Allen wrench into end of adjusting screw. Then, with setting lever at position "9," turn screw counterclockwise until hitch just starts to raise. Tighten the locknut.

157. DRAFT SENSING VALVE ADJUSTMENT. Unbolt and remove lower link pivot pins, then remove hitch lower links from the brackets. Start engine and rotate mode selector fully counterclockwise to draft

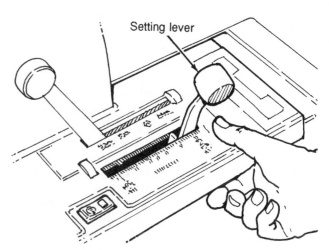

Fig. 194—*Rotate thumb wheels and move both setting lever stops rearward to lock the lever in "0" position.*

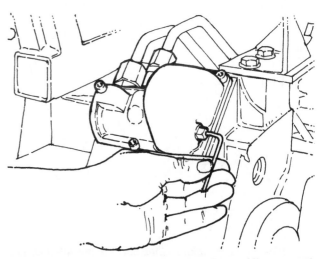

Fig. 195—*Loosen locknut, then use a 3 mm Allen wrench to adjust the position sensing orifice screw.*

(load) control position. Move hitch command switch fully forward into working position detent. Move setting lever rearward to minimum setting position "0."

Refer to Fig. 196, loosen locknuts and use a 5 mm Allen wrench to adjust the two screws. Adjust the left screw and then the right screw until rockshaft arms are fully lowered. Scribe reference marks on pto housing and a rockshaft lift arm. Move setting lever until center of lever is in position "1" on the console. Carefully adjust one of the draft sensing screws until rockshaft arms move and the two reference marks are 19 mm (0.750 in.) apart. Do not exceed this amount. Move setting lever until center of lever is in position "2" on the console. Rockshaft arms will fully lower. Carefully adjust the second draft sensing screw until the two reference marks are again 19 mm (0.750 in.) apart. Then, tighten the two locknuts to a torque of 54-61 N·m (40-45 ft.-lbs.). Reinstall hitch lower links.

158. HITCH DROP RATE ADJUSTMENT. To adjust the hitch drop time, attach an implement or weight to provide a pressure at the hitch cylinder supply line of 4800 kPa (700 psi). Start engine and operate at 2200 rpm. Move setting lever fully forward to position "10." Turn mode control knob clockwise to the position mode. Turn the drop rate control fully clockwise to the fast setting. Move the hitch command switch into the neutral center position. Then, move the command switch to the fast lower (nondetented) position and note the hitch drop time. The time noted should be 1.8-2.0 seconds.

If hitch drop time is not correct, lower the implement or weight to the ground and stop the engine. Loosen locknut (24—Fig. 188) and using a 3 mm Allen wrench, remove adjusting screw (23) completely. Install a 4 mm Allen wrench into internal adjusting plug (14). If drop time is slow, turn drop rate adjusting

plug (14) counterclockwise $\frac{1}{16}$ of a turn. Reinstall the spool centering screw (23) and tighten locknut (24). Recheck and readjust hitch drop time as required to obtain the 1.8-2.0 seconds drop time. After drop time is correctly adjusted, adjust spool centering screw as outlined in paragraph 152. Lower the hitch and stop the engine.

ROCKSHAFT AND LIFT CYLINDER

All Models

159. R&R AND OVERHAUL. To remove the rockshaft and lift cylinder, drain the transmission oil and remove the rear pto housing as outlined in paragraph 166. Remove the rockshaft lubrication hose (10—Fig. 197). Remove the six bolts (27) and remove end cap (26). Discard "O" ring (25). Remove piston (17) and discard seal rings (18 through 21). Use a hammer handle to bump top end of cylinder sleeve loose and slide cylinder sleeve (22) from pto housing. Remove the snap ring from left end of rockshaft (4). Check to see that timing marks are applied to rockshaft and lift arm, then remove left lift arm. Remove right lift arm in the same manner. Remove the two set screws (7) from rockshaft arm (3). Remove the Allen screws and remove the position control housing, then remove position sensing connecting rod from arm in sensing housing. See Fig. 190. Remove rockshaft out left side of housing. Disconnect position sensing rod from rockshaft arm (3—Fig. 197). Remove cotter pin (14), washer (15) and pin (13), then remove rockshaft arm (3) from the housing. Remove piston rod (16) from housing. Remove and discard rockshaft oil seals (1 and 6).

Use new "O" rings, seal rings and oil seals and reassemble in reverse order of disassembly, keeping the following points in mind. Align timing marks on rockshaft arm and lift arms with marks on rockshaft. Lubricate socket in rockshaft arm with special lubricant part number 331-58. Install the second Allen set screw (7) to a torque of 90-107 N·m (66-79 ft.-lbs.). Install oil seals (1 and 6) with lips to outside. Tighten cap retaining bolts (27) to a torque of 125-150 N·m (93-112 ft.-lbs.). Apply a bead of Loctite 515 sealant to face of pto housing. Install pto housing and tighten the bolts to 37-43 N·m (27-32 ft.-lbs.). Fill transmission with Hy-Tran Plus oil to proper level on dipstick.

POSITION CONTROL SENSING VALVE

All Models

160. R&R AND OVERHAUL. With hitch lowered and engine stopped, disconnect the two hydraulic

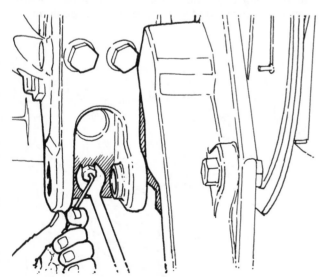

Fig. 196—Loosen locknut, then use a 5 mm Allen wrench to adjust the two (one each side) draft sensing adjusting screws.

tubes from the position sensor housing. Unbolt and remove the sensor housing assembly. Disconnect the sensing connecting rod from the lever in the housing. Remove the left tube adapter, then remove the sensing check valve. Separate the two halves of the sensor check valve. Using a 4 mm Allen wrench, remove the valve return spring and screw. Remove check valve from valve body. Refer to Fig. 198 and remove end plug. Drive out roll pin and remove the arm. Withdraw the check valve actuator spool through end plug opening.

Clean and inspect all parts and renew as required. Use all new "O" rings and reassemble in reverse order of disassembly. The cut-out on the actuator spool must face the check valve. Apply Loctite 242 to the threads of the end plug. Install and tighten the plug. Install arm and secure with roll pin. Install the check valve into the check valve body. Install the screw and return spring assembly and tighten the screw. Screw the two halves of the check valve together, then tighten to a torque of 37-41 N•m (25-30 ft.-lbs.). Install the check valve assembly and the left tube

adapter. Install sensing connecting rod onto the arm. Apply Loctite 515 to face of sensor housing and pto housing. Use new gasket and install sensor housing. Tighten retaining screws to a torque of 11-13 N•m

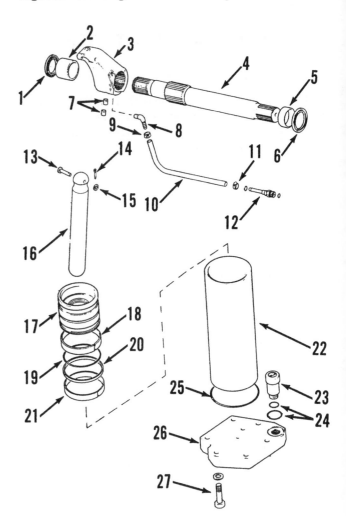

Fig. 197—Exploded view of rear hitch rockshaft and lift cylinder.

1. Oil seal		15. Washer
2. Bushing (L.H.)		16. Piston rod
3. Rockshaft lift arm		17. Piston
4. Rockshaft		18. Wear ring
5. Bushing (R.H.)		19. "O" ring
6. Oil seal		20. Seal ring
7. Set screws		21. Wear ring
8. Elbow		22. Cylinder sleeve
9. Clamp		23. Relief valve
10. Rockshaft lube hose		24. "O" rings
11. Clamp		25. "O" ring
12. Adapter		26. Cylinder cap
13. Pin		27. Bolt (6)
14. Cotter pin		

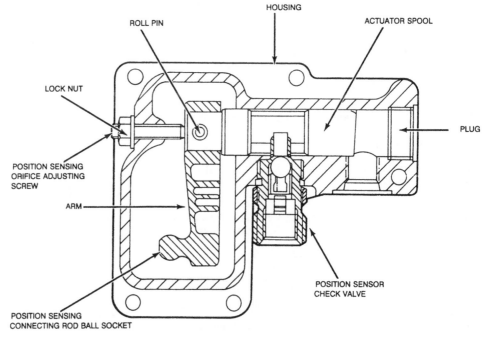

Fig. 198—Cross-sectional view of position control sensing valve.

(8-10 ft.-lbs.). Check and adjust hitch maximum height as outlined in paragraph 153. Adjust position sensing orifice as outlined in paragraph 156.

FRONT THREE-POINT HITCH

All models may be equipped with an optional 3-point front hitch. The hitch is operated by the third remote valve. Refer to paragraph 167 for service procedure on the remote control valve.

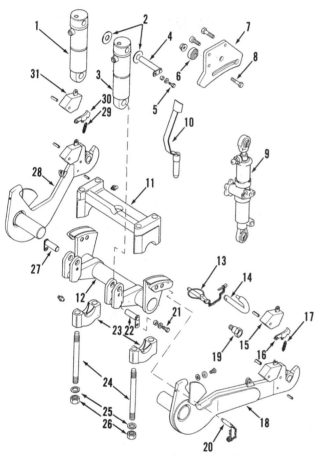

Fig. 199—Exploded view of the optional 3-point front hitch.

1. Lift cylinder (L.H.)	
2. Shims	17. Spring
3. Lift cylinder (R.H.)	18. Lower link (R.H.)
4. Cylinder upper pins	19. Storage position holder
5. Bolt	20. Bolt assy.
6. Stop plate	21. Bolt
7. Bias plate	22. Cylinder lower pivot pin
8. Bolt	23. Support bearings
9. Third link	24. Studs
10. Storage pin (2)	25. Washers
11. Bearing bracket	26. Nuts
12. Hitch rockshaft	27. Cylinder lower pivot pin
13. Linchpin	28. Lower link (L.H.)
14. Third link pin	29. Spring
15. Claw latch support	30. Claw latch
16. Claw latch	31. Claw latch support

All Models So Equipped

161. REMOVE AND REINSTALL FRONT HITCH. To remove the front hitch, disconnect battery cables, raise the hood and remove side panels and front grille. Disconnect and cap the cylinder hoses. Refer to Fig. 199 and remove storage pin (10) from each side. Lower both lower links (18 and 28). Raise right lower lift link (18) to position shown in Fig. 200, then remove the link by sliding it towards the front wheel. Remove left lower lift link (28—Fig. 199) in the same manner. Remove retaining bolts (21) from cylinder lower pivot pins (22 and 27) and remove the pins. Place a floor jack under the hitch rockshaft (12). Remove the nuts and washers (25 and 26) retaining the support bearings (23). Remove the support bearings and lower the rockshaft and bearing bracket (11) to the floor. Remove bolts (5) from the cylinder upper pivot pins (4). Remove pivot pins and shim washers (2) and remove lift cylinders (1 and 3).

Inspect all parts and renew as necessary. Individual parts are not available for the lift cylinders, they are serviced as complete assemblies.

Reinstall front hitch by reversing the removal procedure. When installing support bearing nuts (26), tighten nuts to a torque of 900 N·m (664 ft.-lbs.).

REMOTE CONTROL VALVES

All tractors may be equipped with one to four remote valve circuits. Priority flow goes to the first remote valve, then on to the second, third and fourth valves (if equipped). The third remote valve is used to operate the front 3-point hitch if so equipped. Each remote coupler has an identification plate. The couplers are identified as number 1, number 2, number 3 and number 4, corresponding to the remote control levers. Hoses should be attached so that pushing control levers forward lowers the implement and pulling levers rearward raises the implement. Hydraulic motors should be driven by the first (priority)

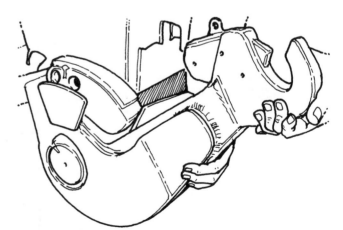

Fig. 200—Raise right lift link of front hitch to position shown when removing.

remote valve. The return oil from all implements must be returned through the hydraulic couplers on remote valves as the return oil supplies pressure lubrication to the transmission. Flow at remote circuits should be 57 L/m (15 U.S. gpm) at 13,790 kPa (2000 psi).

All Models So Equipped

162. R&R AND OVERHAUL. To remove the remote valves, first thoroughly clean exterior of remote valves and surrounding area. Remove retaining nuts from remote valve end cover, then remove the end cover. Disconnect the control valve linkage and remove the remote valves.

Place valves on a clean work bench. Remove the 2 mm Allen set screw and the flow control knob. Remove the shifting lever. Remove the hydraulic couplings from the remote valve and remove the valve number plate. Loosen compression nut (70—Fig. 201 or 202) on flow control valve plug body. Remove the flow control valve plug assembly from the valve body. Remove the adjusting rod (63), thick bearing race (64), bearing (65) and thin bearing race (66) from plug body (68). Remove flow control sleeve (73) with spool (74).

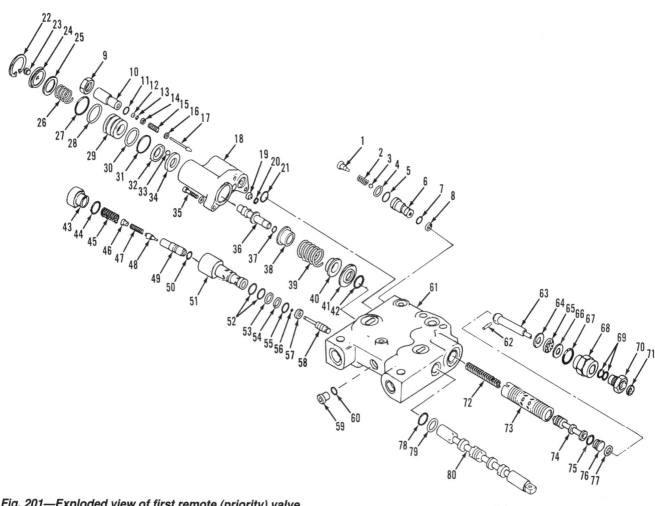

Fig. 201—Exploded view of first remote (priority) valve.

1. Plug	15. Spring	29. Piston	42. "O" ring	55. "O" ring	67. "O" ring
2. Spring	16. Washer	30. Backup ring	43. Cap	56. Snap ring	68. Adjusting plug
3. Ball	17. Pilot relief poppet	31. "O" ring	44. "O" ring	57. Washer	69. "O" ring
4. Backup ring	18. Detent housing	32. Washer	45. Spring	58. Piston	70. Compression nut
5. "O" ring	19. Washer orifice	33. Ball	46. Retainer	59. Plug	71. Seal
6. Check valve body	20. "O" ring	34. Washer	47. Spring	60. "O" ring	72. Spring
7. "O" ring	21. "O" ring	35. Screw	48. Pilot poppet	61. Valve body	73. Sleeve
8. Backup ring	22. Snap ring	36. Detent stud	49. Poppet	62. Roll pin	74. Spool
9. Nut	23. Plug	37. "O" ring	50. "O" ring	63. Adjusting rod	75. "O" ring
10. Adjusting plug	24. Retainer	38. Retainer	51. Check valve body	64. Bearing race	76. Plug
11. "O" ring	25. Spacer	39. Spring	52. "O" rings	(thick)	77. Retaining ring
12. Backup ring	26. Spring	40. Retainer	53. Backup ring	65. Thrust bearing	78. "O" ring
13. "O" ring	27. "O" ring	41. Retainer	54. Backup ring	66. Bearing race	79. Wiper
14. Washer	28. Backup ring			(thin)	80. Main spool

NOTE: Only the first remote valve has two rows of holes in sleeve (73) and center land on spool (74).

Remove snap ring (77), plug (76) with "O" ring (75) and spool (74) from sleeve (73). Remove flow control valve spring (72). Remove check valve assembly (43 through 54), then remove check valve piston assembly (56 through 58). Remove snap ring (56) and dashpot washer (57) from piston (58). Remove check valve

end cap (43) with "O" ring (44), spring (45) and main poppet assembly. Remove spring retainer (46), spring (47) and pilot poppet (48) from main poppet (49).

Place remote valve body in a press and push the detent spring retainer (24—Figs. 201, 202 or 203) down 0.25 mm (0.010 in.) and remove snap ring (22) from detent housing. Release pressure and remove snap ring (22), retainer (24) with dust plug (23) and

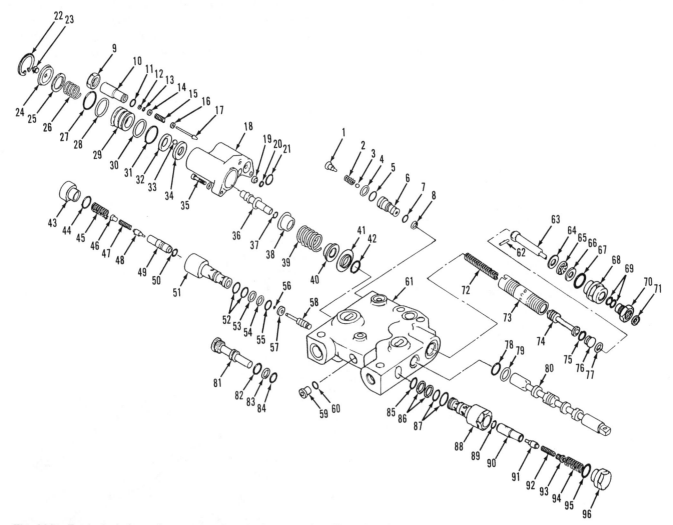

Fig. 202—Exploded view of second or fourth remote valve. Note the absence of a center land on spool (74) and the single row of holes in sleeve (73).

1. Plug		34. Washer	50. "O" ring	65. Thrust bearing	81. Plug
2. Spring	18. Detent housing	35. Screw	51. Check valve body	66. Bearing race	82. "O" ring
3. Ball	19. Washer orifice	36. Detent stud	52. "O" rings	(thin)	83. Backup ring
4. Backup ring	20. "O" ring	37. "O" ring	53. Backup ring	67. "O" ring	84. "O" ring
5. "O" ring	21. "O" ring	38. Retainer	54. Backup ring	68. Adjusting plug	85. "O" ring
6. Check valve body	22. Snap ring	39. Spring	55. "O" ring	69. "O" ring	86. Backup ring
7. "O" ring	23. Plug	40. Retainer	56. Snap ring	70. Compression nut	87. "O" rings
8. Backup ring	24. Retainer	41. Retainer	57. Washer	71. Seal	88. Check valve body
9. Nut	25. Spacer	42. "O" ring	58. Piston	72. Spring	89. "O" ring
10. Adjusting plug	26. Spring	43. Cap	59. Plug	73. Sleeve	90. Poppet
11. "O" ring	27. "O" ring	44. "O" ring	60. "O" ring	74. Spool	91. Pilot poppet
12. Backup ring	28. Backup ring	45. Spring	61. Valve body	75. "O" ring	92. Spring
13. "O" ring	29. Piston	46. Retainer	62. Roll pin	76. Plug	93. Spring retainer
14. Washer	30. Backup ring	47. Spring	63. Adjusting rod	77. Retaining ring	94. Spring
15. Spring	31. "O" ring	48. Pilot poppet	64. Bearing race	78. "O" ring	95. "O" ring
16. Washer	32. Washer	49. Poppet	(thick)	79. Wiper	96. Plug
17. Pilot relief poppet	33. Ball			80.	

detent spring (26). Remove the three retaining screws (35) and remove the detent housing (18). Remove "O" ring (21), washer orifice (19) and "O" ring (20), then remove spacer (25) and detent piston (29) with "O" rings (27 and 31) and backup rings (28 and 30) from housing (18). Remove the outer detent ball washer (32), eight detent balls (33) and inner detent washer (34). Loosen locking nut (9) on lever kickout regulator adjusting plug (10). Using a 6 mm Allen wrench, remove adjusting plug from detent housing (18). Remove the poppet valve assembly (12 through 17) from adjusting plug. Separate backup ring (12), "O" ring (13), washer (14), spring (15) and washer (16) from relief poppet (17). Remove the main spool assembly from the valve body (61). Remove "O" ring (42) and retainer (41) from spool (80) and remove "O" rings (78) and wiper (79) from body (61). Using a 5 mm Allen wrench, remove detent stud (36)

from main spool (80). Remove outer spring retainer (38), spring (39) and inner spring retainer (41).

Using a 3 mm Allen wrench, loosen plug (1) in end of check valve body (6) and remove the check valve assembly. Remove the plug (1), spring (2) and ball (3) from check valve body (6). Remove "O" rings (5 and 7) and backup rings (4 and 8) from check valve body.

Clean and inspect all parts and renew any showing excessive wear or other damage. Check the free length of springs against the following new spring specifications:

Spool centering spring
(39—Figs. 201, 202 or 203)
Free length . . . . 60.05 mm
(2.364 in.)

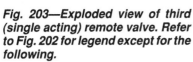

Fig. 203—Exploded view of third (single acting) remote valve. Refer to Fig. 202 for legend except for the following.

- 1A. "O" ring
- 38A. Spool stop
- 81. Plug
- 82. "O" ring
- 83. Backup ring
- 84. "O" ring
- 85. "O" ring
- 86. Backup ring
- 87. "O" ring
- 88. Check valve body
- 89. "O" ring
- 90. Poppet
- 91. Pilot poppet
- 92. Spring
- 93. Spring retainer
- 94. Spring
- 95. "O" ring
- 96. Plug
- 97. Flow control spring (outer)
- 98. Flow control spring (inner)
- 99. Pressure spool
- 100. "O" ring
- 101. Backup spring
- 102. "O" ring
- 103. Plug

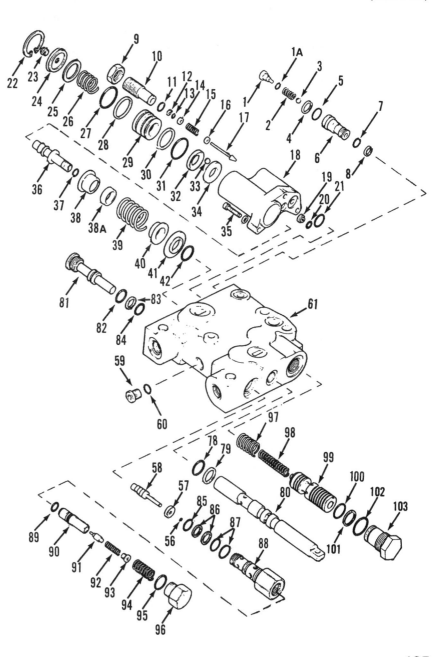

Flow control outer spring
  (97—Fig. 203)
    Free length . . . . 54.33 mm
                                            (2.138 in.)
Flow control inner spring
  (98—Fig. 203)
    Free length . . . . 67.71 N·m
                                            (2.665 in.)

Renew all "O" rings and backup rings, and reassemble by reversing the disassembly procedure, keeping the following points in mind. Tighten check valve body (6—Figs. 201, 202 or 203) to a torque of 16-19 N·m (12-14 ft.-lbs.). Tighten plug (1) to 4-5 N·m (3-4 ft.-lbs.). Apply Loctite 242 to threads of detent stud (36) and tighten to 12-16 N·m (9-12 ft.-lbs.). Tighten detent housing retaining screws (35) to 5-7 N·m (4.5-6 ft.-lbs.). Apply Loctite 242 to threads of spring retainer (46) and tighten to 4-5 N·m (3-4 ft.-lbs.). Tighten end cap (43) to a torque of 34-41 N·m (25-30 ft.-lbs.). Tighten check valve body (51) to torque of 34-41 N·m (25-30 ft.-lbs.). Tighten adjusting plug (68) to a torque of 41-47 N·m (30-35 ft.-lbs.).

Tighten compression nut (70) until a torque of 0.9-1.1 N·m (8-10 in.-lbs.) is required to rotate adjusting rod (63).

When reinstalling the remote valves, use all new "O" rings, and tighten 10 mm nuts to 35-40 N·m (26-30 ft.-lbs.) and 12 mm nuts to 61-69 N·m (45-51 ft.-lbs.).

163. ADJUST DETENT KICK-OUT PRESSURE. Connect a flowmeter to remote valve couplings. Operate tractor to heat hydraulic oil to a temperature of 50° C (120° F). Set engine speed at 2200 rpm. Turn flow control knob to maximum position. Turn load valve on flowmeter fully open. Move remote valve control lever fully rearward in detent position. Observe the pressure gauge on flowmeter and slowly close the load valve. Detent kick-out pressure should be 15,170 kPa (2200 psi). If not, loosen locknut (9—Figs. 201, 202 or 203) and turn adjusting plug (10) as required to obtain the correct setting. Tighten locknut to a torque of 36-42 N·m (27-31 ft.-lbs.).

REAR POWER TAKE-OFF

All tractors are available with a two-speed (540 or 1000 rpm) independent power take-off. The pto is controlled by a lever on the console that actuates a cable-operated hydraulic control valve. The control valve controls the engagement and disengagement of the hydraulically operated, multiple-disc pto clutch. The pto is driven by the input shaft coupled to the lubrication and charge pump drive shaft (Fig. 176). Pto speeds are based on engine speeds of 1877 rpm for the 540 pto speed and 2210 rpm for the 1000 pto speed.

LINKAGE ADJUSTMENT

All Models

164. To adjust the pto control valve linkage, move the pto control lever into the disengaged position. Refer to Fig. 204 and remove pin (2). Make sure the control valve lever is in the disengaged position. Adjust the cable adjusting nuts (1) so that pin (2) can be installed freely. Tighten adjusting nuts.

PTO NEUTRAL START SWITCH

All Models

165. Disconnect the pto neutral start switch connector and connect a multimeter to terminals "A" and "B" of the switch. With pto control lever in disengaged position, adjust nuts (3—Fig. 205) until the multimeter shows continuity, then tighten adjusting nuts.

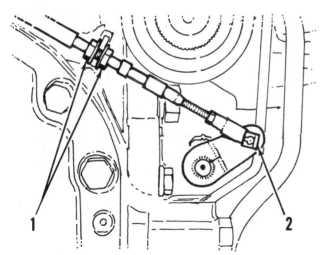

Fig. 204—View showing adjustment points of pto control valve linkage.

1. Adjusting nuts
2. Pin

Engage the pto and check to see there is no continuity in engaged position.

PTO HOUSING

All Models

166. REMOVE AND REINSTALL. To remove the pto housing, drain oil from transmission and pto housings. Disconnect and remove 3-point hitch linkage. Remove retainer pins and remove drawbar. Disconnect pto control cable. Disconnect pto speed sensing switch cable. Disconnect rear pto lubrication tube. Disconnect the remote valve control rods. Disconnect and cap the return hose from the remote valve manifold and the pilot and pressure compensator tubes. Disconnect the position control tubes from the hitch position sensor housing. Refer to Fig. 206, remove locknut and sensing bracket stop bolt. Re-

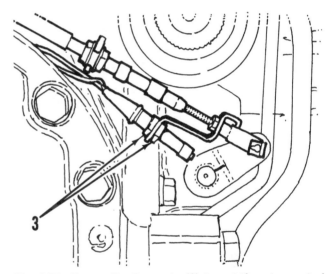

Fig. 205—Turn adjusting nuts (3) to set the pto neutral start switch.

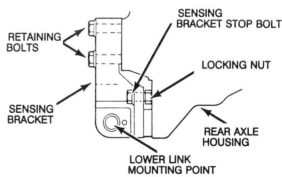

Fig. 206—Load (draft) sensing bracket used on all models. Left bracket is shown. Right bracket is similar.

move retaining bolts and lift off left lower link sensing bracket. Disconnect the hydraulic lift cylinder tube and the pto valve inlet tube and fitting. Attach a hoist to the unit, remove retaining bolts and lift assembly from tractor. Remove pto drive shaft.

Clean faces of pto housing and rear main frame. Apply Loctite 515 to the face of rear main frame and install the pto assembly. Tighten mounting bolts to a torque of 195-250 N·m (145-185 ft.-lbs.). Complete the installation of the pto housing in reverse order of removal procedure, keeping the following points in mind. Use new "O" rings. Tighten lift cylinder tube bolts to a torque of 27-31 N·m (20-23 ft.-lbs.). Tighten lower link sensing bracket mounting bolts to 610-730 N·m (450-540 ft.-lbs.). Refer to Fig. 206 and install the stop bolt and locking nut. Turn the stop bolt

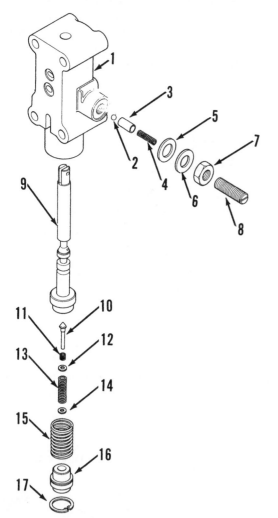

Fig. 207—Exploded view of pto control valve.

1. Valve body	
2. Detent ball	
3. Guide	
4. Spring	10. Valve
5. Seal washer	11. Secondary spring
6. Washer	12. Washer
7. Locknut	13. Primary spring
8. Adjusting screw	14. Washer
9. Spool	15. Return spring
	16. End cap
	17. Snap ring

clockwise into the axle housing until the head of the bolt contacts the sensing bracket. Then, turn the stop bolt counterclockwise one-half turn and tighten the locking nut to a torque of 650-730 N·m (480-540 ft.-lbs.).

NOTE: Tighten locking nut against the axle housing, NOT the sensing bracket.

Refill transmission with Hy-Tran Plus oil to correct level on dipstick. Adjust pto control valve linkage as in paragraph 164 and adjust lower link sensing screws as outlined in paragraph 157.

CONTROL VALVE

All Models

167. R&R AND OVERHAUL. With the pto housing assembly removed as outlined in paragraph 166, place the assembly on a work bench. Unbolt and remove the oil baffle plate. Remove the pto valve retaining bolts and lift off the valve assembly.

Refer to Fig. 207 and loosen locknut (7). Then, counting the turns, unscrew and remove adjusting screw (8). Record the number of turns and discard the seal washer (5). Remove the detent spring (4), guide (3) and ball (2) from valve body (1). Remove snap ring (17), end cap (16), return spring (15), washers (12 and 14), springs (11 and 13) and valve (10). Slide spool (9) from valve body.

Clean and inspect all parts and renew any showing excessive wear or other damage. Spool (9) and valve body (1) are not available as separate parts. Reassemble by reversing the disassembly procedure. Use a new seal washer (5) when installing adjusting screw (8). Install screw (8) the same number of turns recorded during disassembly. Tighten locknut (7).

Reinstall control valve in reverse order of removal and using new "O" rings. Tighten control valve retaining bolts to a torque of 73-87 N·m (54-64 ft.-lbs.). Apply a bead of Loctite 515 to pto housing and install baffle plate. Tighten bolts to 37-43 N·m (27-32 ft.-lbs.).

Reinstall pto housing as in paragraph 166.

NOTE: The detent assembly holds the valve in engaged position when the engine is operating. When engine is stopped and the hydraulic pressure is lowered, the detent releases the spool to disengaged position.

PTO CLUTCH

All Models

168. R&R AND OVERHAUL. To remove the pto clutch, first remove the pto housing assembly as outlined in paragraph 166. Place pto assembly on a work bench, then unbolt and remove the front seal retainer and discard the seal. Unbolt and remove the oil baffle plate. Remove the clutch hub (28—Fig. 208). Remove thrust bearing (26) and thrust washer (25) from inside the hub (28) and seal ring (27) from outside of hub. Remove bolt (3), washer (2) and two "O" rings from center of clutch hosing (11), then remove clutch assembly.

To disassemble the pto clutch, remove snap ring (23—Fig. 208) and backing plate (22). Remove the four friction discs (21) and four separator plates (20). Place clutch assembly in a press and using spring compressor tool (CAS-1992) or equivalent, compress piston return spring (17). Remove snap ring (19), then remove clutch from press. Remove spring retainer (18) and piston return spring (17). Bump clutch housing against a wooden block to remove piston (12). Remove "O" rings (14 and 15) and seal rings (13 and 16) from piston. Remove snap ring (4), brake disc (5), brake plate (6), wave spring (7) and second brake disc (10). Remove the three pins (8) with "O" rings (9). Remove needle roller bearing (24) from clutch housing (11).

Clean and inspect all parts and renew any showing excessive wear or other damage. Renew friction discs (21) if the thickness is less than 2.4 mm (0.094 in.). Renew separator plates if thickness is less than 2.1 mm (0.082 in.).

Use all new "O" rings and seal rings and reassemble in reverse order of disassembly. Soak friction discs (21) in clean Hy-Tran Plus oil before installing. Tighten clutch retaining bolt (3) to a torque of 101-113 N·m (75-83 ft.-lbs.).

Apply Loctite 515 to the baffle plate contact surface of the pto housing. Install baffle plate and tighten bolts to a torque of 37-43 N·m (27-32 ft.-lbs.). Install a new oil seal in front seal retainer. Apply Loctite 515 to seal retainer surface and install retainer. Tighten bolts to a torque of 37-43 N·m (27-32 ft.-lbs.).

Refer to paragraph 166 and reinstall pto assembly.

PTO GEARS AND SHAFTS

All Models

169. R&R AND OVERHAUL. Remove the pto housing assembly as outlined in paragraph 166. Then, remove the pto control valve as in paragraph 167 and the pto clutch as in paragraph 168.

Refer to Fig. 209 and remove cotter pin (7) and speed sensing rod (8). Remove snap ring (19) and withdraw pto output shaft (20). Remove and discard oil seal (10). Remove snap ring (11). Remove snap ring (25) and spring (23) from sensing piston (22). Remove front bearing snap ring (9). Drive output shaft sleeve (14) rearward to remove bearing cup (12). Drive sleeve (14) forward to separate the two driven gears

(15 and 18). Install a 10-24 × 2-inch UNC Allen head screw with a minimum thread length of 1.5 inches into shift collar pin (17). Turn screw clockwise until it contacts the sensing piston and remove the collar pin. Rotate shift collar (16) 180° and remove the second collar pin. Remove the sensing piston (22) from the outer sleeve (14) and discard "O" ring (21). Remove the outer sleeve (14) and bearing cone (13) from rear of housing. Remove the 540 rpm gear (18), shift collar (16) and 1000 rpm gear (15). Remove bearing cone (5) and bearing cup (6). Remove snap ring (33) and use a slide hammer puller to remove input shaft (30) from front of housing. Use a blind hole puller to remove rear bearing cup (27). Remove seal rings (26), then remove bearing cones (28 and 31) from input shaft (30).

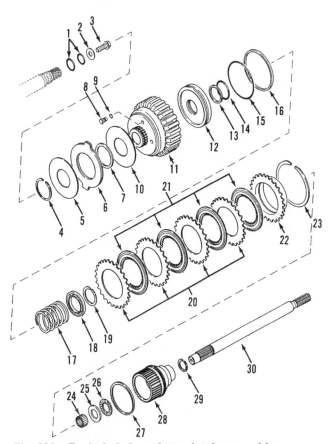

Fig. 208—Exploded view of pto clutch assembly.

1. "O" rings	16. Seal ring
2. Washer	17. Piston return spring
3. Bolt	18. Retainer
4. Retaining ring	19. Snap ring
5. Brake disc	20. Separator plates (4)
6. Brake plate	21. Friction discs (4)
7. Wave spring	22. Backing plate
8. Pin	23. Snap ring
9. "O" ring	24. Needle roller bearing
10. Brake disc	25. Thrust washer
11. Clutch housing	26. Thrust bearing
12. Piston	27. Seal ring
13. Seal ring	28. Clutch hub
14. "O" ring	29. Snap ring
15. "O" ring	30. Drive shaft

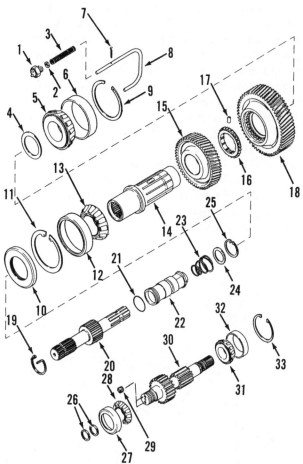

Fig. 209—Exploded view of rear pto gears and shafts and related parts.

1. Switch	18. Driven gear (540 rpm)
2. Washer	19. Snap ring
3. Spring	20. Reversible output shaft
4. Washer	21. "O" ring
5. Bearing cone	22. Inner sleeve
6. Bearing cup	(sensing piston)
7. Cotter pin	23. Spring
8. Speed sensing rod	24. Washer
9. Snap ring	25. Snap ring
10. Oil seal	26. Seal rings
11. Snap ring	27. Bearing cup
12. Bearing cup	28. Bearing cone
13. Bearing cone	29. Plug (2)
14. Outer sleeve	30. Input shaft
15. Driven gear (1000 rpm)	31. Bearing cone
16. Collar	32. Bearing cup
17. Shift collar pin (2)	33. Snap ring

Clean and inspect all parts and renew any showing excessive wear or other damage. Use all new "O" rings and seal rings during reassembly.

To reassemble, lubricate and install new seal rings (26—Fig. 209) on rear of input shaft (30). Use a seal compressor tool (CAS-2005-3) to compress the seals. Using a bushing driver, install bearing cup (27). Install bearing cones (28 and 31) on input shaft. Install input shaft and front bearing cup (32). Install snap ring (33). Attach a slide hammer to front of shaft, then move shaft forward and rearward to seat the bearings. Use a dial indicator to measure shaft end play. End play must be 0.025-0.127 mm (0.001-0.005 in.). If not, select a snap ring of correct thickness to provide correct end play. Snap rings are available in various thicknesses. Install front bearing cup and cone (5 and 6). Install 1000 rpm gear (15), shift collar (16) and 540 rpm gear (18). Align the holes in shift collar (16) with slots in outer sleeve (14) and install sleeve assembly from rear of housing through the two gears and shift collar. Install new "O" rings (21) onto sensing piston (22) and install sensing piston into outer sleeve (14). Separate the 540 and 1000 rpm driven gears. Slide shift collar on the outer sleeve spline until hole in shift collar aligns with the slot in the outer sleeve (14) and groove on sensing piston (22). Install shift collar pin (17) until it seats into the groove of the sensing piston. Rotate the collar 180° and install the second pin.

NOTE: The shift collar must slide freely with no binding within the slots of the outer sleeve. If binding occurs, raise pins slightly.

Install rear bearing cup (12) and the thinnest snap ring (11) available. Rotate the shaft to seat the bearings and using a dial indicator, check end play of outer sleeve (14). End play must measure 0.025-0.127 mm (0.001-0.005 in.). Snap rings (11) are available in various thicknesses. Select and install the snap ring that provides the correct end play. Install a new oil seal (10) into the housing. Install the reversible pto output shaft (20) and snap ring (19). Install speed sensing rod (8), spring (3) and cotter pin (7).

Reinstall pto clutch and control valve as outlined in paragraphs 168 and 167, then reinstall pto assembly as outlined in paragraph 166.

CAB AND SEAT

TRACTOR CAB

All Models So Equipped

170. REMOVE AND REINSTALL. To remove the cab, first block the front wheels to prevent tractor from moving. Place wooden wedges between front axle and front bolster to prevent tractor from tipping. Disconnect cables from battery negative terminals. Position a suitable stand, such as tool number CAS 10853, under the transmission housing. Raise rear of tractor and remove the rear wheels to provide working clearance for cab removal. Raise the hood and remove side panels. Note that early models have a manually operated hood release lever located on the left side of the instrument console, while later models utilize an automatic hood latch. Remove the exhaust extension tube and the rear hood panel.

On right side of tractor, disconnect brake reservoir filler hose and return hose from brake reservoir (2—Fig. 210). Disconnect air conditioner self-sealing coupling (3). On models so equipped, disconnect hood release cable and remove cable from support bracket. Disconnect steering tube (4) from steering hose and remove from support bracket. Remove oil cooler tube support bracket. Unbolt and remove rear hood bracket. Close heater valve. Place a 20 liter (5 gal.) container beneath heater hoses, disconnect hoses and drain coolant.

On left side of tractor, disconnect throttle cable from injection pump and remove cable from support brackets. Disconnect electrical harness connectors from cab bulkhead connectors. Disconnect air conditioner hose at self-sealing coupling. Disconnect steering hose and remove from support bracket. Disconnect pto control cable (5—Fig. 211).

At rear of tractor, disconnect breather tube from fuel tank. Disconnect parking brake cable (6—Fig. 212) and remove cable from support bracket and clamp. Disconnect inching cable (7) and remove from support bracket. Disconnect remote valve actuating rods. Remove trailer electrical connector from remote valves. Disconnect right-hand console wiring harness connector located on underside of cab. Disconnect and cap the brake tubes. Disconnect transmission range selector rods and speed selector rods. Disconnect mechanical front-wheel-drive clutch control valve hydraulic hose (if equipped). Disconnect draft control hydraulic tubes and remove tube support clamp. Disconnect park brake cable from brake lever. Mark the steering hoses to ensure correct reassembly, then disconnect the hoses between compensator valve and steering hand pump. Disconnect creeper control cable

Fig. 211—Pto control cable (5) is located on left side of tractor at the rear of rear axle housing.

Fig. 210—Disconnect points on right side of tractor for cab removal.

1. Hood latch	3. Air conditioner coupling
2. Brake reservoir	4. Steering tube

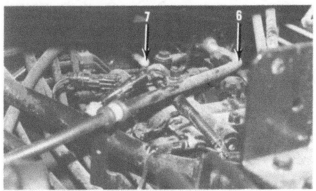

Fig. 212—View of parking brake cable at rear of tractor.

(if equipped). Disconnect steering return hose at hydraulic oil filter base. Disconnect wiring harness electrical connector from Powershift module (if equipped).

To ensure that cab is removed safely, it is recommended that a cab lifting frame be used when removing cab. Lifting frame may be fabricated using dimensions shown in Fig. 213. Remove plugs from rear of cab frame and attach lifting eyes (3—Fig. 214) to cab using 12 mm (grade 10.9) bolts. Nylon slings (2) may be used to attach front of cab lifting frame to cab hand grips. Attach lifting frame to cab as shown in Fig. 214. Disconnect cab ground strap. Remove cab front and rear mounting nuts. Using a suitable overhead hoist, slowly raise cab from tractor while making sure all harnesses, cables and hoses have been disconnected and are not binding.

To reinstall cab, reverse the removal procedure. Tighten cab mounting nuts to 265 N·m (195 ft.-lbs.).

OPERATOR'S SEAT

Mechanical Suspension

171. R&R AND OVERHAUL. To remove seat assembly, remove eight mounting screws and remove seat from tractor. Unbolt and remove seat from suspension assembly.

To disassemble suspension, remove retaining plugs (10—Fig. 215) attaching rubber boot (9) to upper and lower housing. Move boot down away from lower housing (12) and insert a block of wood between upper and lower housings to prevent suspension from collapsing. Drive roll pin from weight adjustment rod

(23). Turn weight adjustment knob (7) counterclockwise to release tension on weight adjustment spring, then drive roll pin from knob and remove knob and thrust bearing (6). Remove rubber boot from housings. Remove weight adjustment springs (21) and the weight adjustment rod (23) and bar (22). Remove retaining shaft (11) from lower housing, then separate upper and lower housings. Remove suspension arm retaining shaft (11) and separate suspension arm assembly (5) from upper housing (8). Drive roll pins from cam follower bearings (14 and 17) and remove retaining clips from bushings (19). Remove the

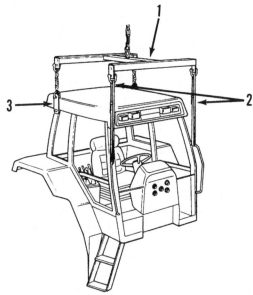

Fig. 214—Cab lifting bracket (1) is attached to cab using slings (2) through the hand grips and lifting eyes (3) at rear of cab.

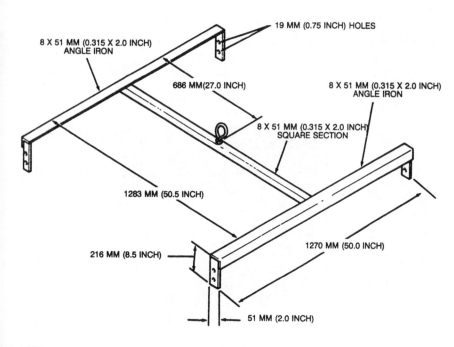

Fig. 213—Cab lifting bracket can be fabricated using the dimensions shown in this illustration.

mounting arm (18) and cam assembly (16) from upper housing.

Inspect all parts for wear or damage and renew as necessary. To reassemble, reverse the disassembly procedure.

OPERATOR'S SEAT

Air Suspension

172. R&R AND OVERHAUL. To remove seat assembly, first disconnect cables from battery negative terminals. Disconnect seat wiring harness connector located under the cab. Remove seat mounting screws and remove seat assembly. Unbolt and remove seat from suspension assembly.

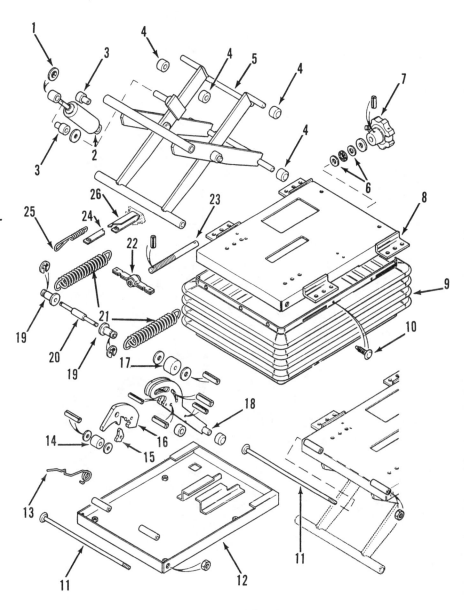

Fig. 215—Exploded view of mechanical suspension seat components.

1. Retaining clip
2. Shock absorber
3. Bushing
4. Rollers
5. Suspension arm assy.
6. Thrust bearing
7. Weight adjustment knob
8. Upper housing
9. Rubber boot
10. Retaining plug
11. Retaining shaft
12. Lower housing
13. Spring
14. Bearing
15. Latch
16. Cam
17. Bearing
18. Mounting arm
19. Bushings
20. Shaft
21. Springs
22. Weight adjustment bar
23. Weight adjustment rod
24. Weight adjustment tape
25. Spring

To disassemble suspension, unbolt and remove seat mounting plate and insulator. Remove retaining plugs (24—Fig. 216) from upper housing (1) and lower housing (25). Remove knob (21) from air control switch (20). Remove rubber boot (23) from suspension assembly. Remove retaining shaft (17) from lower housing, and push grommet and wiring harness through opening in lower housing. Remove cap screw attaching the air spring (16) to lower housing, then separate lower housing from upper housing. Remove air control switch from upper housing. Remove nuts and bolts retaining shock absorber mounting bracket and control lever block to upper housing. Separate suspension assembly from upper housing. Remove shock absorber (12), fore and aft control lever block (4), air spring (16) and air compressor (18) from suspension frame (14).

Inspect all parts for wear or damage and renew as necessary. The air compressor assembly is serviced as a complete unit. To reassemble, reverse the disassembly procedure.

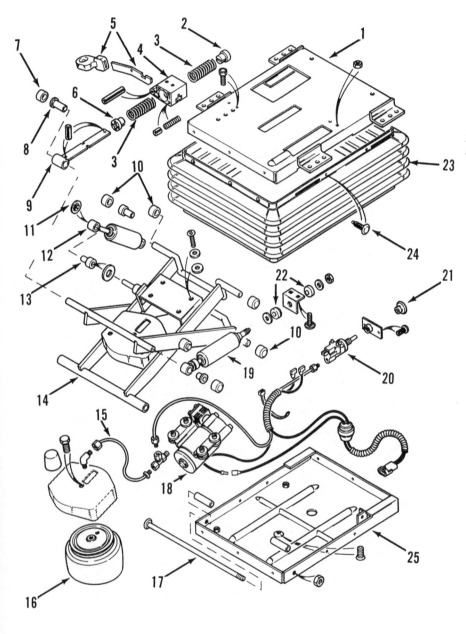

Fig. 216—Exploded view of air suspension seat components.

1. Upper housing
2. Spacer
3. Spring
4. Control lever block
5. Fore & aft control lever
6. Spacer
7. Roller
8. Bearing
9. Bar
10. Rollers
11. Push nut
12. Shock absorber
13. Bushing
14. Suspension arm assy.
15. Air tube
16. Air spring
17. Retaining shaft
18. Air compressor
19. Shock absorber
20. Air control switch
21. Knob
22. Bushings
23. Rubber boot
24. Retaining plug
25. Lower housing

NOTES

NOTES

NOTES

NOTES

NOTES

NOTES